ROC PUBLIC LIBRARY
3 1
D0856083

asrt
National
Library
Partnership
asrt
NATIONAL RADIOLOGIC TECHNOLOGY WEEK 2019
WAVES
of the
FUTURE
NOV. 3-9, 2019
asrt.org/NRTW

SEX, LIES, & BRAIN SCANS

'One is tempted to pick up *Sex, Lies, & Brain Scans* simply for its enticing title. However, once picked up, this book is not easy to put down. This volume provides an introduction to advances in brain science related to the most human of foibles including prejudice, lying, impulsive decision-making, and lapses in moral behaviour. It introduces the reader to insights arising predominately from the application of functional magnetic resonance imaging (fMRI). It highlights the power of this technique to illuminate previously hidden aspects of human brain function, essentially generating data that speaks in some ways to the neural codes that the brain uses to generate complex behaviours. It also highlights challenges involved in this research that complicate the interpretation of fMRI data. Lastly, the book highlights new, and potentially questionable, applications of fMRI, including its potential use as a lie detector or as an adjunct to the effective marketing of commercial products. *Sex, Lies, & Brain Scans* is written by Professor Barbara Sahakian of the University of Cambridge, a leading figure in this area of research, and her graduate student, Ms Julia Gottwald. It provides an important introduction to breakthroughs emerging from neuroimaging for people who are wondering what all the recent fuss regarding the brain is all about.'

JOHN H. KRYSTAL, MD, ROBERT L. MCNEIL, JR.
Professor of Translational Research and Chair,
Department of Psychiatry,
Yale University

BARBARA J. SAHAKIAN
JULIA GOTTWALD

SEX, LIES, & BRAIN SCANS

HOW fMRI REVEALS WHAT REALLY GOES ON IN OUR MINDS

OXFORD
UNIVERSITY PRESS

Great Clarendon Street, Oxford, OX2 6DP,
United Kingdom

Oxford University Press is a department of the University of Oxford.
It furthers the University's objective of excellence in research, scholarship,
and education by publishing worldwide. Oxford is a registered trade mark of
Oxford University Press in the UK and in certain other countries

First Edition published in 2017
Impression: 2

Published in the United States of America by Oxford University Press
198 Madison Avenue, New York, NY 10016, United States of America

British Library Cataloguing in Publication Data
Data available

Library of Congress Control Number: 2016948328

ISBN 978–0–19–875288–2

Printed in Great Britian by
Clays Ltd, St Ives plc

I dedicate this book to friends and family and especially to Trevor, Jacqueline, and Miranda Robbins and Richard Sahakian, who have always inspired and encouraged me.

Barbara J. Sahakian

To my family, who shower me with love and support.
To my friends, whose wonderful weirdness I need.
To Carl, who makes life truly brilliant.

Julia Gottwald

ACKNOWLEDGEMENTS

We thank Trevor Robbins for his insightful comments on the final draft of this book, and Hermann Hauser, Cristian Regep, and Martin Ducker for helpful discussion. We also thank our colleagues and students, who have strived to understand the brain in health and disease, and the volunteers who have generously given their time and support to research. Finally, we thank the Vice Chancellor of the University of Cambridge and the Communications and Festival Staff for their strong commitment to engaging the public in science.

CONTENTS

LIST OF FIGURES

LIST OF PLATES

1

HOW DOES NEUROSCIENCE IMPACT SOCIETY?

What if you could be much smarter? Imagine getting a brain scan and seeing which of your brain circuits do not work at their absolute optimum. Would you target them with drugs or electrical stimulation to reach your maximum potential—bespoke cognitive enhancement? How about finding the love of your life or the perfect job matched to your neural signature? Or what if we could screen for terrorists using a mind-reading device? One day, these scenarios might become possible thanks to rapidly progressing techniques and sophisticated methods of data analysis.

One of the most influential techniques in neuroscience is functional magnetic resonance imaging (fMRI). Every day we read about new and exciting neuroimaging studies. Neuroscientists all over the world use fMRI and other techniques to see the brain in action. These studies shed light on our feelings, thoughts, and behaviours. And by learning more about the circuits and chemicals in the healthy brain, we can also develop new treatments for neurological and psychiatric disorders. What is possible thanks to neuroimaging today was unimaginable only fifty years ago. Prior to these new techniques, neuroscientists had to rely on much more limited ways to learn about the nervous system. Often, the only way to explore the anatomy of the human brain was to wait until somebody died. We have gained amazing insights from these post-mortem studies, but they only progressed slowly and could not teach us much about the function of brain areas. We often forget how neuroimaging has revolutionized the field. For the first time, we could observe the living human brain.

Neuroimaging methods provide valuable information that, in some cases, cannot be obtained by other means. In 2005, Adrian Owen, John Pickard, and colleagues carried out a groundbreaking study at the University of Cambridge, the MRC Cognition and Brain Sciences Unit, and the University of Liège. They tested a 23-year-old woman who was involved in a traffic accident. She suffered a severe brain injury and her assessment five months after the accident concluded that she was in a vegetative state. The researchers scanned the brain of the woman and asked her to imagine playing tennis. Surprisingly, areas involved in motor control became active in her brain. When she was then asked to imagine visiting all the rooms of her house, areas related to spatial navigation were activated. This pattern of brain activation was the same in healthy control subjects, so the woman must have had conscious awareness.[1] This awareness was not picked up in a classical clinical assessment, but the researchers were able to detect it using fMRI.

Martin Monti from the MRC Cognition and Brain Sciences Unit and Audrey Vanhaudenhuyse from the University of Liège took these findings one step further. They wanted to see if they could use these brain activation patterns to communicate with patients who fulfilled clinical criteria of the vegetative state. One out of fifty-four tested patients was able to learn a technique that used his brain activation to answer yes and no questions. When he wanted to answer a question with 'yes', he was instructed to imagine playing tennis. For 'no', he imagined navigating through a city or his home. The researchers were indeed able to use this technique to decode the patient's answers.[2]

These are remarkable and exciting findings. The patients had a traumatic brain injury and a specialized clinical team concluded that they were lacking awareness. However, the fMRI results clearly show that these patients were aware and able to communicate with the outside world. Such an exceptional way of communication could be used to assess if the patient is in pain and if he has any wishes. fMRI might enable us to re-establish communication with a small proportion of patients who were thought to be unaware.

Findings like this illustrate how fundamentally neuroimaging can change our understanding of thoughts, motivations, and behaviour. Neuroscientific methods such as fMRI enable us to investigate the neural circuits involved in a certain process. But we can also address questions in very new ways. We can investigate processes that participants are not even aware of—entering the subconscious. They might be eager to hide something, such as a lie or a racist world view, which could certainly affect their behavioural measures. These are the limits of psychological research, which could be overcome by neuroimaging.

Exciting findings like these spark media interest. Good science journalists draw conclusions that are based on scientific evidence and inform us about these studies in a balanced way. Sometimes, however, findings can be taken out of context or over-interpreted, partly because scientists themselves are not communicating their findings clearly enough. Neuroimaging studies are very complex and sometimes hard to understand. In this book, we will describe some of the most exciting fMRI studies and what they really show. Is mind-reading possible thanks to fMRI? Can a brain scan show if you lie? Can we peek into your brain and know what you will buy?

Think of newspaper headlines as a blinking billboard. Behind the billboard, you have mechanics working tirelessly to keep the circuits working and to make new connections—these are the scientists. Usually, the lights in the front represent the circuits well, but sometimes there are small defects. In this book, we aim to give you a tour of the mechanics behind the billboard. We want to show you how scientific studies are conducted and what the mechanics actually see: the connections they make. Importantly, these connections are constantly updated and changed. Science is dynamic. New findings contribute to our knowledge and sometimes override old theories. And sometimes we are not sure yet which explanation of the finding is the best, because there is evidence to support different explanations.

This will not be a comprehensive account. Rather, this is a whistle-stop tour through some of the most exciting scientific studies and their possible implications. Many of these studies raise ethical questions, because they open up new ways to understand and predict human behaviour. We have to ensure that we utilize neuroscience for the benefit of society and that unintended consequences do not cause harm. We will have to decide where we need regulations and which applications of neuroscience we are keen to promote. Our aim is to provide information about the science and to open the field up for discussion. Neuroethical questions cannot be answered by scientists alone. Finding these answers needs to be a collaborative effort of neuroscientists, ethicists, policymakers, and the public. This book will not provide definite answers to these ethical questions. We simply want to equip readers with the basic scientific background to come to an informed opinion.

First, you will need to understand what fMRI does. This neuroimaging technique is used to measure and map brain activity. It is based on the same technology as MRI—magnetic resonance imaging. Most of us know somebody who has had to undergo an MRI scan for medical reasons. For example, if your child hits his head on the playground and starts vomiting soon afterwards, doctors might want to check for brain bleeding or swelling using MRI. The technique can also be used to detect tumours, and examine internal organs or joints. Almost any part of the body can be examined with MRI. The focus here is on hydrogen. This chemical element is one of the main building blocks of your body—it is part of the water molecule, fat tissues, and proteins. In its core—the nucleus—there is a single proton.

When a combination of a strong magnetic field and radio waves is applied to these hydrogen protons, they are forced to leave their natural position. Think about these protons like the needle of a compass. When you put a strong magnet near the compass, the needle will point towards this magnet. This is what happens when you are moved into the scanner—the protons align. Now imagine that you use your finger to move the needle. Like your finger, the radio wave from the MRI machine forces the needle out of its position. When you remove your finger, the needle will return to its previous position. Similarly, after the radio waves are switched off, the protons return to their previous position. By doing so, they emit a signal—another radio wave. This signal can be detected by the MRI machine. Importantly, hydrogen protons behave slightly differently in different tissues, and the time to return to their previous position depends on the surrounding tissue. Therefore, the signals from different tissues can be differentiated. For example, we can see clear differences between fat tissue and bones on the MRI scan and detect anatomical abnormalities.

In contrast, functional MRI tracks changes related to blood flow. Your brain is constantly active, even when you sleep. This activity changes depending on what you do. If you start a new task, the brain areas involved in this process will alter their activity. If a brain area becomes more active, it needs energy, and so more oxygenated blood will flow in this region. There are differences between oxygenated and deoxygenated blood—one of them is their magnetic property. Essentially, the fMRI machine picks up these magnetic differences to give us an indirect measure of the neural activity. Figure 1 shows an fMRI scanner—the same machine can be used for structural and functional MRI. In fact, researchers usually perform a structural MRI of the head

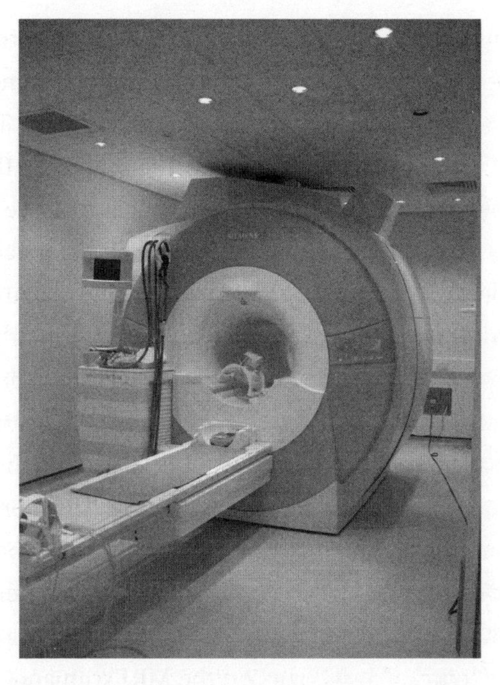

Fig. 1. A magnetic resonance imaging scanner at the University of Cambridge, which was used for some of the studies mentioned in this book.

before they start the fMRI experiment, so that they can map the brain activity to the anatomical structure.

Many fMRI experiments have a block design. For example, you would do a certain task for thirty seconds and then have a rest for thirty seconds. Researchers can then compare your brain activation during performance of the task and the rest period, and so find out which brain areas were involved in doing the task. A different approach is event-related fMRI. Rather than blocks, the participant is presented with single events, usually in a random order. For example, you would see a pleasant image, an unpleasant image, and then two pleasant images, rather than a whole block of pleasant and a block of unpleasant images. Both designs have their advantages and disadvantages. Block designs often miss the element of surprise—after seeing five pleasant pictures in one block, you have clear expectations about the sixth one. In contrast, event-related designs allow the researchers to present different types of stimuli in a random order. However, by focusing on single events rather than blocks, the researchers lose statistical power. The brain's response to a series of stimuli could be stronger and more straightforward to analyse than its response to a single stimulus.[3] In the end, which design is more suitable depends on the experimental question.

That is the basic principle of fMRI, but like all techniques it has limitations. The data generated are complex and require rigorous analysis by trained researchers with sophisticated software. The brain contains about 86 billion neurons[4] which are organized in different areas and networks. We cannot measure the activity of single neurons with fMRI, but we can break the brain down into smaller cubes—so-called voxels. The voxel size depends on the technology which is used for the experiment, but usually a voxel measures a few millimetres in each of the three dimensions. A voxel can contain about a million neurons, depending on its size and the brain region involved.

The signal we detect stems from a range of neurons and their activity in a given time. Because the human brain contains so many neurons and therefore so many voxels, we detect signals from a large number of voxels. These signals are computed and analysed statistically. Slight differences in these statistical tests can lead to different results, so rigorous and appropriate analysis is critical.

The spatial resolution of fMRI is relatively good, which means that we can locate a brain signal quite well. However, the temporal resolution is relatively poor, because there is a delay between neural activity and blood flow. The analysis tries to take this into account, but this limitation means that some very quick neural events cannot be detected.

Another limitation is that we can rarely establish causality. For example, if we see abnormal activation in a brain scan and the scanned person also shows abnormal behaviour, we can rarely ever tell which one causes the other. It might be that the abnormal brain activation started earlier and the behaviour followed as a consequence; or performing such an abnormal behaviour could itself change the brain activation. The relationship between brain activation and behaviour is a correlational one and cannot provide direct evidence of causation.

Moreover, most neuroimaging studies are based on groups rather than on individuals. Because there are so many differences between people, we need to scan a range of participants to find out which patterns are generally present. Often, researchers compare two groups of people, for example a control group and a group of patients. This mode of analysis means that we can rarely draw conclusions about the individual, because the analysis focuses on the average activation of the groups. Say a group of thirty subjects shows a certain activation in a given region. Now you scan one more subject but you do not find activation in this region of his or her brain. Is this necessarily concerning? No, because our individual brain structure and function can be slightly different.

These limitations are important to keep in mind when interpreting fMRI studies. Even though the technology is often seen as objective and 'hard scientific proof', the method is not error-proof. Findings have to be interpreted critically within a context. Colourful images of brain activation maps look compelling, but we cannot simply accept them as visual proof. Research has shown that publications which include brain images score higher on perceived scientific reasoning than ones featuring bar graphs or no images, even if the brain scans do not contain any additional information[5] (although a later study did not replicate this effect, but did find that the addition of neuroscientific language can make a bad explanation more convincing[6]). People like 'hard evidence' and a brain picture (or the inclusion of neuroscientific language) seems to serve that purpose. But no matter how beautiful and persuasive a brain image, ultimately it cannot explain everything, and it is certainly not without flaws.

A discussion of the brain invariably involves some use of the standard terminology for brain regions. There is a logic behind

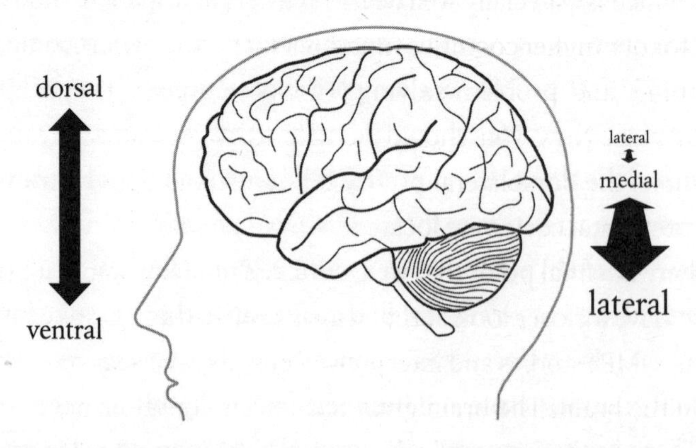

Fig. 2. Illustration of a brain showing the dorsal–ventral and lateral–medial axes.

these names. Once you apply this logic, you will be able to locate many brain regions by their names. The words you will need are shown in Figure 2:

- dorsal (from Latin 'dorsum'—meaning 'back'): the brain regions which are on the side of your spine when you look up
- ventral (from Latin 'venter'—meaning 'belly'): the brain regions on the side of your belly when you look up
- lateral (from Latin 'latus'—meaning 'side'): the regions towards the left and right side of your brain
- medial (from Latin 'medius'—meaning 'middle'): the brain regions towards the midline between your forehead and chin

Finally, the outer layer of the brain is called the cerebral cortex, or simply the cortex, and it will be mentioned in connection with many imaging studies in this book. The frontal cortex is an area which is especially well developed in humans and is important for our higher cognitive functions such as decision-making, planning, and problem-solving. This is the basic ABC of the brain map. Now you should be able to work out where, for example, the dorsolateral prefrontal cortex and the ventromedial prefrontal cortex are located.

There is a final point that we would like to make before we go on to have a closer look at the neuroimaging studies. When we discuss fMRI studies and interpret the results, we focus on activity in the brain. The brain gives rise to our consciousness, our personality, our sense of self. Your mind is just your brain in action: it is not a separate entity. When we observe brain activity, we observe the mind. Therefore, it does not make sense to justify our behaviour with phrases like, 'My brain made me do it'. Mind and brain are intimately linked. There can be a brain without a mind (when the person is dead), but no mind without an active brain. Think about it like your heart. The organ itself can exist on its own just like the brain can exist without electrical activity. But the heartbeat cannot exist without a working heart, just like the mind cannot have a life of its own independently of the brain. Accepting this relationship is crucial to our understanding of human behaviour.

Genes as well as structural and functional changes in the brain can create conditions for vulnerability or resilience. What insights can we gain from neuroimaging studies about what motivates and drives us, why we make the decisions and choices we do, and whether we can predict our behaviour?

Now that we have covered the basics, let us dive into the scientific studies and what they have revealed so far.

2

CAN NEUROSCIENTISTS READ YOUR MIND?

Every one of us is a mind-reader. We do it every day, and our ability to communicate and cooperate with our social group depends on it. What does my boss think of me? Is my partner happy? Is my son going to run after his football and into the busy street? You might not be aware of it, but you are processing a great deal of social information whenever you interact with another person. Your brain works like a detective conducting an investigation about another person's thoughts and mental state, based on evidence from, for example, facial expression, body language, tone of voice, and your previous knowledge of the person. These factors form a bigger picture and most of us are quite skilled at coming to the right conclusion—reading the mind.

We only become aware of our investigative skills when we observe people who lack them. Social cognition is greatly impaired in some psychiatric conditions such as autism spectrum disorder. The majority of patients have difficulty understanding their peers' thoughts and feelings. They read facial cues less accurately,[1] have trouble putting themselves into another person's shoes,[2] and are less able to form relationships. This shows how important social cognition is for our everyday lives.

Even though we are skilled mind-readers, we are far from perfect. Mistakes are common and some people are impressively capable of hiding their true feelings. Those in some professions have to be particularly good at hiding emotions, such as politicians, actors, and poker players. What if there was an objective, accurate, scientific method of reading someone's mind? Such a scenario could be both dream and nightmare: though undoubtedly useful, in the wrong hands such technology holds the potential for abuse.

How Much Mind-Reading Is Really Possible Today?

Some scientific studies have tested the possibility of mind-reading, and this has attracted much press attention. Newspaper headlines claimed that we now have a 'brain scan that can read people's intention',[3] or 'So our minds CAN be read: magnetic scanner produces these actual images from inside people's brains.'[4] But what is the scientific evidence?

Many studies try to predict the actions of participants. Subjects perform tasks while their brain activity is recorded by an fMRI machine. The scientists often use an approach called machine learning, in which a computer does not need a pre-programmed solution to a problem. The model learns from the data fed into it from a training set and improves over time to make predictions about new data. This approach is highly useful, because it makes the model flexible and driven by the data, rather than the hypothesis of the researcher. Machine learning technology is already used in many areas, such as the optimization of online search engines or self-driving cars. But it is also a valuable tool for research.

In the early stages of mind-reading experiments, John-Dylan Haynes and colleagues from the Max Planck Institute for Human Cognitive and Brain Sciences were able to predict whether you intended to add or to subtract two numbers presented to you[5] or whether you were going to press a left or right button.[6] While this may not make a Hollywood thriller, it was a remarkable achievement around ten years ago. Since then, the techniques and computer power have improved, allowing the decoding of more elaborate processes.

One basic approach to 'eavesdrop on one's thoughts' is to find out which nouns a person is thinking about. In 2008, Tom Mitchell and his colleagues at Carnegie Mellon University reported a breakthrough in the field of mental state decoding.[7] They invited nine participants to an fMRI experiment in which each of them was presented with sixty different nouns. The subjects had to imagine the properties of every word when it was presented, for example they saw the word 'castle' and might imagine 'knight', 'cold', and 'stone'. Their patterns of brain activity were fed into a computer model, for all nouns apart from two. The model then predicted the pattern of activity for the two excluded words, based on what it had learned from the other fifty-eight patterns. Afterwards, it was given the two scans that had been left out and matched them with the two words. On average, the computer was right more than seven times out of ten. If that seems too easy, the model was then trained with fifty-nine out of sixty words, presented with a new activity pattern and asked to predict the matching word—this time, out of a pool of 1001 nouns. The model performed equally well in this scenario. Impressive, isn't it? Bear in mind, though, that this does not make the computer a perfect mind-reader. It needs a lot of training, and in about a third of the cases it was still wrong. And our thoughts are a lot more complex than just single words. In order to decode a mental state, we need more sophisticated techniques.

Jack Gallant's group from the University of California has developed a highly impressive method to reconstruct a film clip that a subject was watching purely based on the fMRI recordings.[8] The subjects—in this case, three researchers who are also authors of the paper—first watched a set of film trailers known to the computer. It was therefore possible to associate the trailers with brain activation patterns. After this training stage, the subjects then watched a second set of clips, but this time the computer had no information about the content of the film trailers. Instead, the model used the known associations to form a reconstruction of the clip. The reconstructions were blurry and not very detailed, as you can see in Plate 1. Can you guess what the original trailers showed? Read the figure caption to find out. Brain signals are still too complex and the fMRI technology is not capable of capturing very rapid neurotransmission. While there are considerable advances required, it is nonetheless remarkable what we are able to achieve at present. Technology is constantly evolving and this might be just the beginning.

One possible application of reconstructing film clips is to reconstruct our 'inner films': our dreams and memories. A Japanese research team led by Tomoyasu Horikawa tried its luck with the former in 2013. They scanned the brain activity of three people while they were falling asleep and entering the dreaming stage, then woke them up and asked for a description of the dreams. The subjects had to repeat this more than 200 times to give the researchers a good pool of data. Among the dreams were ordinary experiences ('I saw a scene in which I ate or saw yoghurt'), but also some unusual scenarios ('I saw something like a bronze statue [...] on a small hill'). Key words were assigned to different categories (for example, 'food' or 'geological formations'). Subjects were then shown photos from each of these categories and again their brain activity was measured. These data were fed into the model. The computer compared the activation of seeing an image while awake with the activation during the dream and made a prediction as to whether a certain content was present in the dream or not ('Did the person dream about food?'). The model does not work perfectly, but it is reasonably accurate: on average three out of five times.[9] One important weakness is the lack of objectivity. Participants had to describe their dream contents to the researchers and these reports form an important data set for the study. But who knows how accurately we remember our own dreams? To date, there is no objective way of measuring this, so we have to rely on subjective reports. Nonetheless, the study illustrates an exciting way of analysing internal representations.

As we have discovered, neuroscientists are able to make good predictions about simple and straightforward thoughts. But this leaves out a big and important component of our mental state—our emotions. When your boss says he is happy with your work, a thought identification device may confirm that it is actually your work he is thinking about (and not his afternoon golf game). But does that tell us that he is indeed happy? We need a different type of information to be sure: a peek at his emotions. Karim Kassam and his colleagues from Carnegie Mellon University have taken that peek successfully, using actors from the local community for their study. The actors were asked to put themselves into nine emotional states (anger, disgust, envy, fear, happiness, lust, pride, sadness, and shame) while in the scanner. They accomplished the experience of these emotions by imagining scenarios they developed before the scan. Rather than pretending, they were asked to actively immerse themselves into the feelings in more than a hundred trials, presented in a random order.

While the emotionally drained actors recovered, their data were fed into a computer model which could learn and improve its assessments with experience. The model was able to identify the correct emotions of a subject on average four out of five times when comparing the scan patterns with previous trials of the same subject. That is already remarkable, but here comes an amazing twist: the model was still correct on average seven out of ten times when comparing the neural activity of one subject with the scans from *other* individuals. Thus emotions seem to have a similar neural basis among individuals (or at least among different actors). This seems to be more true for emotions like anger and less so for shame, but overall the model predicted all nine emotions with impressive accuracy.[10]

The Limitations

While these recent advances are certainly remarkable, they only work within strict limits. There is no 'one-size-fits-all' approach, especially for complex thoughts, because our brains are like every other part of the body: they vary between individuals. Your neural representation of eating a yoghurt (especially while being ashamed) might be different from that of your neighbour. To date, computers cannot handle this variability without being trained. The participants in the described studies underwent functional imaging for extended periods of time; they saw a large number of video clips or were awoken from their dreams annoyingly often. This initial training stage is important for the model to adapt to your personal activation patterns—to 'get to know you'. Therefore, it is also impossible today to use these techniques 'undercover'—you would certainly know if you spent hours in a big noisy fMRI machine, repeatedly drifting in and out of sleep.

Moreover, the widespread use for potential mind-reading would not be practical at present. The equipment is expensive, heavy, and not portable. It requires a special isolated room, trained personnel, and complex analyses. You have to lie still in an MRI scanner for a relatively long period of time before researchers or clinicians can get a good scan of your brain. This is a challenge for young children, but some adults also struggle to lie still for any length of time—especially those with motor symptoms such as attention deficit hyperactivity disorder (ADHD) patients. Could we have much faster machines where only a part of the head needs to be immersed in the scanner? Techniques such as magnetoencephalography (MEG) make this possible. In contrast to fMRI, MEG measures the magnetic field produced by the brain rather than blood flow. The magnetic fields change quickly and the technique is able to detect these rapid changes. In fact, the temporal resolution of MEG is considerably higher than in fMRI. But nothing comes without a price: tracing the brain activity back to a precise location is much harder in MEG. Researchers sometimes combine these techniques to get the best from both, but this approach is not very practical, since it requires multiple scans and the integration of big data sets. We are still looking for the 'holy grail' of neuroimaging: a temporally and spatially precise method that is also cheap, safe, easy, and portable.

There are some ideas about how to overcome at least the issue of portability. Some newer techniques, including diffuse optical tomography (DOT), which uses light rather than magnets, are in development. The accuracy of these new systems seems to be catching up with the current gold standard in the field—fMRI.[11]

Reading personal thoughts in great detail using fMRI is science fiction for now. The experimental conditions have to be tightly controlled and the context well defined. The computer predictions reflect blurry shapes of a seen film or the presence/absence of food in your dream, to name a few. Plate 2 shows Adrian Owen, who was involved in the remarkable work on patients in the vegetative state that we discussed in Chapter 1. Next to him is a scan of his brain, taken while he was lying in an fMRI scanner performing a particular task. He asked the world's top neuroimagers to identify what he was doing. Here are their answers:

1. Remembering something
2. Tracking a stimulus on the screen
3. Shifting attention from one thing to another
4. Deciding which of two responses to make
5. Doing sudoku
6. Switching attention
7. Tapping a finger in response to a stimulus
8. Counting
9. Looking at disgusting pictures
10. Nothing

What was Adrian Owen really doing? He was telling a lie! Not a single expert could identify the correct activity. They clearly did not lack creativity, but reading someone's mind from a brain scan without any information about the context is currently impossible. Mind-reading does not work in isolation and even the most sophisticated machine has to get to know the subject first. But the field is moving ahead rapidly. Advances in machine learning techniques and new imaging methods could overcome the current limitations of mind-reading. In the future it might be possible to know what your favourite politician is really thinking, no matter how good his acting skills are.

Good vs Bad Mind-Reading?

Most researchers working on technological refinements such as the development of new imaging techniques or better computer modelling have worthy, ethical aims. Better technology would make brain scans safer, easier, and cheaper. They could be used to advance our understanding of the brain and inner thought processes. In some cases, they could also enable us to communicate with people who have lost their ability to speak, due to being in a coma or to muteness. However, there is also the potential for abuse, especially if a large number of people suddenly have access to this technique. When do we have the right to keep our thoughts private? When does the government, an organization, or an individual have the right to read our thoughts? Is it ethical to use mind-reading techniques, for example, at an airport to screen for terrorists? Many people are already uncomfortable with full-body scans: what if a machine had access to your thoughts and emotions?

Clear ethical guidelines for applications of brain scans and potential mind-reading will be needed. We as a society will have to debate where we want to draw the line between responsible, beneficial usage of the techniques and what constitutes their abuse.

3

A RACIAL BIAS HIDING IN YOUR MIND?

Would you call yourself racially biased? Nowadays, most people would answer this question with 'no', but it depends how one defines this bias. There are two kinds of racial bias we need to consider: explicit and implicit bias. Explicit racial bias is a conscious, intentional belief that all members of one race are characterized by specific qualities which make them inferior or superior to other races. This bias is sometimes called 'old-fashioned racism'. Based on self-report questionnaires it seems that this type of racism is declining in the general population.[1] Most constitutions have declarations of equality, and many societies do not tolerate explicit racism. However, this growing political and social pressure might only affect outspoken, open racism that people admit to. It does not necessarily mean that the individual agrees with society's values personally.

Moreover, people might not consider themselves racists, but does that mean they treat people of all races equally? An implicit racial bias is a prejudice against a given race that is more unconscious and not necessarily something the person is aware of. It can affect behaviour and make one act in a racist fashion, or it can be controlled to avoid behaviour that we know to be morally or socially wrong. In this chapter, we will argue that any racial bias, even if it is subtle and implicit, has the potential for disadvantaging people solely based on their race. It can affect the delivery of health care, chances for job applicants, or court rulings, to name only a few examples. Therefore, it is important to be aware of these biases. They turn out to be more common than you might think.

Measuring the Hidden Racial Bias with the Help of Neuroscience

In a study in 2000 led by Elizabeth Phelps from New York University, participants were given a questionnaire called the Modern Racism Scale,[2] which explores racial attitudes. For statements like, 'Over the past few years, the government and news media have shown more respect to blacks than they deserve', the (all white) participants reported to what extent they agreed or disagreed. As expected, they reported very low self-perceived racial bias. However, their behaviour and brain activity told a different story. The researchers did not simply rely on the participants' self-report: they tested racial attitudes implicitly. The participants were shown pictures of either black or white male faces while they were in the fMRI scanner. Subjects did not know the people shown to them: the faces were taken from a college yearbook and they all had neutral facial expressions.

The task was described to the participants as a memory experiment (in fact, the Modern Racism Scale was given to them after the scanning, so they did not know that this study was actually investigating racial bias). The subjects saw one face at a time and had to decide whether it was the same face as the last one or a different one. Afterwards, the researchers looked at the brain activation and compared the signals for black and white faces.[3] One region that they were particularly interested in was the amygdala, which is implicated in threat processing and the detection of socially important stimuli, among other functions. The amygdala is located deep inside the brain, behind the eye and towards the centre of the brain. It is shaped like an almond, hence 'amygdala', derived from the Greek word for the nut.

The amygdala activation was correlated with implicit scores of racial bias, measured with the Implicit Association Test (IAT). This test is presented on a computer and it consists of several stages. Imagine that you are a participant in the study. In the important stages of the task, you see two pairings on the screen. For example, you would see 'black/bad' in the top left corner and 'white/good' in the top right corner. Then, words are presented in the centre of the screen and you have to decide whether you categorize them as 'black' or 'white' (for example, the first name 'Temeka') or as 'good' or 'bad' (like the word 'wonderful') using a button press. The categories are then reversed—'black/good' and 'white/bad'. Figure 3 shows an example of IAT trials.

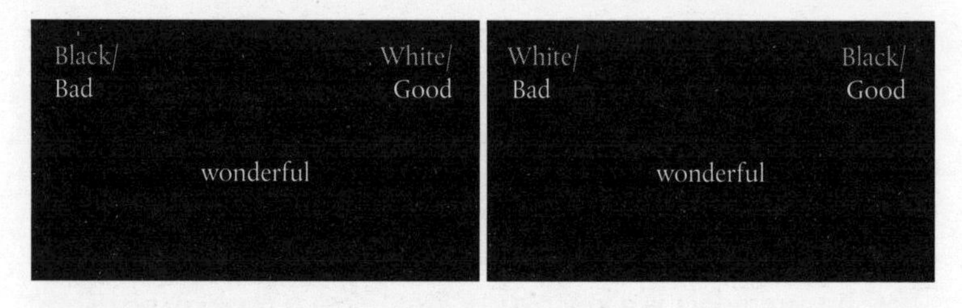

Fig. 3. Example of two trials on the Race Implicit Association Test. If you can put positive words such as 'wonderful' faster into the 'White/Good' than the 'Black/Good' category, you might have an implicit pro-white bias.

When you are done, the researchers look at your reaction time. If you put 'wonderful' into the pleasant category faster when you have the 'white/good' option than with the 'black/good' option, it shows an implicit negative bias against blacks (or a positive bias for whites, depending on how you want to look at it). The reasoning goes like this: if you have a pre-existing association of white people and something good but not the same association with blacks, you find the 'white/good' category much easier to use, while the use of the 'black/good' category feels less intuitive and thus takes longer.[4] If you think that you would never show such a bias, take the test yourself! A collaboration of several US universities provides a free version of the IAT online: (https://implicit.harvard.edu/implicit/takeatest.html).

But now back to the participants in 2000. They showed a longer response time for 'black/good' (and also 'white/bad') in the task and thus a pro-white bias. The observed implicit bias contradicts the attitudes reported on the questionnaires. This is interesting in itself, but the effect was already known back then. The remarkable fact about this study is that the authors decided to compare the imaging data with the IAT results. Their significant finding:

the higher the pro-white bias on the implicit test, the greater the amygdala activation to black vs white faces.[3] You might not consider yourself biased, but it is very possible that your automated responses and brain signals do not support this belief.

What is the psychological basis for racial bias? We have seen that the amygdala is a key region, but there is more evidence to suggest that an out-group bias is a fear-related reaction. Students were asked to complete the IAT in a study led by Sylvia Terbeck and her colleagues from the University of Oxford. As we have seen, the IAT is reasonably good at detecting implicit racial biases (and actually many other prejudices relating to gender, age, weight, sexuality, religion, and other factors, depending on the study design—many of us carry a whole range of social biases). Half the students received a placebo before the test, the other half the beta blocker propranolol, which is used to treat hypertension and anxiety symptoms. The medication did not affect the explicit racial prejudice or mood, but it did significantly reduce the implicit racial bias.[5] Fear and threat-processing seem to play an important role in racial bias.

If the IAT or amygdala activation in response to different faces is not enough to convince you, this next study might. A research team led by Xiaojing Xu in Beijing decided to use a different study design to investigate empathy. They filmed Chinese and Caucasian subjects while they were either touched with a cotton bud or penetrated with a needle in the face. As you can imagine, one stimulus is painful, the other one is not. These videos were then shown to other Chinese and Caucasian subjects.

Empathy is quite a curious thing. Being in pain yourself and watching somebody else in pain activates a very similar network in your brain, but only if you feel empathetic towards that person.[6] The Chinese team wanted to find out whether race has any influence on this measure of empathy. Indeed, both Chinese and Caucasians showed activation of pain-related brain areas when watching somebody of their own race receiving a painful stimulus. This activation was strongly decreased, though, when the person in the video was a member of the other race.[7] Not my race, not my pain?

Can Racial Bias Be Changed?

These studies seem to paint a pessimistic picture of the world and how people of different races interact, but there is hope. We now know that frequent out-group contact improves attitudes by reducing negative biases. A study using a very big data set has shown the remarkable impact of interacting with people of a different group: increased contact reduces prejudice.[8] This idea was first developed in the 1950s by the psychologist Gordon Allport. He hypothesized that this contact has to happen under 'optimal conditions', namely:

(1) in the situation, the interacting groups have to have an equal status;
(2) the groups have common goals (such as winning a football game together);
(3) there is cooperation between the groups;
(4) there is the support of authorities, law, or custom.[9]

In our globalized world, one might think that all these conditions are becoming more and more likely. A follow-up study of the experiment with pain perception and empathy shows some promising results. Again, participants were shown videos of both Chinese and Caucasian subjects receiving painful or non-painful stimuli. However, this time, the team recruited a different type of participant. They only allowed Chinese subjects who had spent a significant part of their childhood outside China to participate. The twenty subjects grew up in the UK, the US, or Canada—that is, in countries where the majority of the population is Caucasian. Strikingly, these people who had lived abroad for so long did not show any significant differences between the two groups: they had equal brain activation patterns in response to Chinese and Caucasian people being in pain.[10] It is yet to be established if these markers of empathetic pain translate into real-life reduction of prejudice and discrimination, but it does look promising. If this proves to be true, should we encourage intergroup contact more strongly? Spending time in a different culture surrounded by people of a different race might be a powerful antidote to racial bias.

The effect of such intercultural experiences might be a more positive attitude towards people of a different race, or it might be a simple matter of familiarity and therefore a decline of the sense of threat. The power of this factor was already shown by the first neuroimaging study of racial bias in 2000 by Elizabeth Phelps and colleagues. What we did not tell you about the study at the beginning of this chapter is that there was a second experiment. Instead of unfamiliar faces, participants were then shown pictures of both black and white celebrities, who were equally famous and of a similar age. Shown were positively regarded people like Muhammad Ali, Denzel Washington, Harrison Ford, and John F. Kennedy. Strikingly, the researchers did not find any of the effects observed with unfamiliar faces. The participants showed no racial bias when looking at celebrities.[3] This could mean that prior knowledge of the person diminishes prejudice.

It turns out that racial bias is more easily modified than one might think. Even something seemingly random like the music people listen to can have a big effect.[11] A research team from the University of Arizona and the University of British Columbia led by Chad Forbes recruited white undergraduate students who showed very low self-perceived racism and high motivation to respond without prejudice. The students were told that the study was about the influence of music on spatial processing. They were shown faces of black and white people and had to report whether they were presented on the left or right side of the screen—a reasonably simple task. During the task, no music, a heavy metal song, or a rap song was played in the background.

The participants showed no difference in amygdala activity (remember, the almond-shaped brain region implicated in threat-processing) in response to white or black faces when listening to no music or a heavy metal song. If you are interested, it was the song 'Only One' by Slipknot, which was previously rated as triggering negative emotions, but not associated with a specific race.

A difference was seen, however, when they listened to a song which was shown to evoke negative feelings and to be more stereotypic of black Americans. This song was 'Straight Outta Compton' by N.W.A. Feel free to listen to it, but be warned that it contains strong, violent, and misogynistic language. This time, participants did show the differential amygdala activation to blacks vs whites. Something as seemingly irrelevant as a song changes amygdala activation. A little frightening, isn't it?

It gets even more concerning. We introduced Allport's contact hypothesis before, which states that out-group contact decreases prejudice. You may have noticed that many of the studies investigating racial bias test this idea in white subjects. How do African Americans feel towards other black people? The sad truth is that many of them also show a greater amygdala activity to black faces than white ones—a pro-white bias.[12] This effect was certainly not seen in all participants: some studies contradict these findings.[13] But in the light of this evidence we cannot argue that a pure lack of contact leads to prejudice. It must be associated with cultural learning as well, which might be harder to overcome.

Why Do We Need to Think about Racial Bias?

Apart from personal opinions and attitudes, why is racial bias important? A number of studies have shown serious effects of this unconscious bias. One particularly convincing experiment was conducted in 2004 by Marianne Bertrand from the University of Chicago, and Sendhil Mullainathan based at the Massachusetts Institute of Technology. They sent out almost 5,000 CVs to employers in Boston and Chicago, some with stereotypical white names like Carrie or Brad, some with names more typical for African Americans, such as Latoya or Jamal.

The researchers were extremely careful to match the CVs for all other variables like years of experience, skills, military experience, and many more details. They ended up with two pools of statistically identical CVs, which only differed in the name of the applicant. They then counted the number of call-backs for each application. Their results were shocking. Fictitious white applicants received 50 per cent more call-backs than African-American ones, given equal qualifications.[14]

Maybe employers have racial biases, but surely doctors are trained to see all people equally? Sorry to disappoint you, but even medical doctors are not error-proof. Physicians attending one of two conferences in 1996 or 1997 were invited to participate in a study led by Kevin Schulman from the Georgetown University Medical Center. The physicians were shown a recorded interview with a patient with chest pain (who was actually just an actor) and their medical file. Afterwards, the doctors were asked to give a diagnosis and to make a recommendation for treatment. A total of 720 doctors, none of whom knew the real purpose of the study, took part. The actors reported similar symptoms and the medical files contained similar information. Nonetheless, physicians were much more likely to recommend an important and useful treatment to whites than blacks (and also more likely to men than women, but this is a different topic).[15] Given the dangers of heart disease, this result is particularly concerning.

Unequal treatment is not restricted to job applications and health care. Marvin Chun's group from Yale University investigated racial bias in the context of court rulings. The (again all white) participants were shown different cases of employment discrimination and told to decide how much money they would award the black victim. The subjects also completed the IAT and underwent fMRI while looking at white or black faces. The scientists were then looking for a potential link between the IAT scores, brain activation, and awarded money. The IAT did not predict the amount of money, but the fMRI signal did. The amount of neural activation in response to black vs white faces was enough to make a prediction of how much money the participant would award the victim.[16] This suggests that fMRI could be a useful tool to detect a racial bias, for example when selecting people for a jury.

Screening for Racial Bias?

It is noteworthy that in the studies described there was great variability between the participants: some people showed stronger implicit racial bias than others. Should we use this information to screen for unbiased people before they are allowed to be in powerful positions? If we can use brain scans to identify such discriminatory biases, would we use them to screen doctors, teachers, or judges? Some people might show an implicit bias in such a test, but would not act upon it because they know that this would be morally and socially wrong. Brain scans and tests of implicit bias cannot determine racism; only behaviour can. Some of us might be very skilled at executing cognitive control and inhibiting ourselves from discriminating against people based on race. How would we measure this?

The approach of screening people does not seem to be justified or feasible. As discussed in Chapter 2, fMRI is expensive and only available within a defined setting. Moreover, even if it were possible to screen people routinely for their racial attitudes in a reliable way, would this be sensible or desirable? How many people without any racial bias would we be able to find? How many qualified, educated, intelligent people would we need to fire? Most of these people are not evil-minded and do not actively believe that one race is superior to another. The old-fashioned racism is dying. Implicit biases are much more common now and many people are not even aware of them. It seems more sensible to actively make people aware of these biases. It might be the case that people did not lie on the tests like the Modern Racism Scale, but really do think that they have equal attitudes towards people regardless of their race. These undetected biases are potentially more dangerous, because they cannot be considered and actively avoided if the person is not even aware of them. Education and positive contact with out-group members could play an important role in diminishing these biases, rather than firing all those who make important decisions about people's lives and show a subconscious bias.

There is hope: the studies with celebrities and with immigrants show that attitudes can change. Also, when participants are asked to focus on the individual rather than the racial group, responses become more positive. Mary Wheeler and Susan Fiske from Princeton University were able to demonstrate this effect. White subjects were shown photos of unfamiliar faces in a scanner, and asked to say either (a) if the person was over or under 21 years old, or (b) if the person would like a certain vegetable. The first task is relatively easy and does not require complicated processing. The second one, however, was designed to force the participant to see the presented face as belonging to an individual rather than just a racial group. The participants had to really focus on the person, the individual traits, and try to make an assumption. Simple stereotypes were not enough (although you might argue that facial details cannot give you much information about whether or not a person will like celery either). The first task showed results similar to the early studies on racial bias: greater amygdala activation for out-group faces. The second one, however, diminished this difference; it was even reversed in one of the brain areas.[17] This suggests that it would be helpful to make an extra effort to see people as individuals before seeing them as members of racial groups.

Would it not be ideal if we could live in a world and interact socially on the merits and personal traits of individuals, rather than on fearful or aggressive 'brain activations' and impulses? How can we use these important findings from neuroscience, which inform us about racial bias, to create a better and fairer society with a greater sense of well-being? Some universities and institutes have diversity training.[18] This training not only aims to address social biases in their explicit form, but may also help people become more aware of implicit biases.

We might also be able to 'unlearn' our biases, possibly even during sleep. In a study conducted at Northwestern University by Xiaoqing Hu and colleagues, participants completed a counter-bias training course. They had to pair black faces with positive words, such as 'pleasure' or 'sunshine'. They also completed a second course, which aimed to reduce a common gender bias: that women are not good at science. Each training course was associated with a specific sound: one for racial and one for gender bias. The training was very effective at reducing the social biases in the short term. After one week, however, the implicit bias returned to baseline.

But the researchers had an idea to counter this: they told the participants to take a nap after the training. When the subject entered the deep-sleep phase, one of the sounds associated with counter-bias training was played to them repeatedly. This presumably reactivated the memory of the training and reinforced the learning effect. Indeed, the bias was even further reduced after the nap when the corresponding sound was played. The other bias did not change. After one week, the sleep-reactivated bias was still lower than at the beginning, while the other one returned to baseline level.[19]

The study leaves many questions unanswered. What happens in the long term? How many training phases and sleep reactivations would we need for a lasting effect? Does the reduced bias have a significant effect on real-life interactions? Future research will show if we are able to 'sleep off' our social biases. If we can, this would be a remarkably cheap and easy way to change people's attitudes. It would undoubtedly be exciting and useful, but would also require new ethical guidelines. We want to avoid a *Brave New World* scenario in which people are manipulated in their sleep, to be controlled by a few powerful leaders. If sleep-training proves to be a powerful technique, we will need mechanisms to regulate its applications. The reduction of social biases is certainly an important goal. Initiatives and methods, like those we have discussed, will be needed to achieve a fair society, where people do not have to hide their gender, race, religion, appearance, etc. to have equal chances.

4

THE PERFECT LIE DETECTOR?

When was the last time you lied? Today? Last week? Last month? You might not be aware of it, but the chances are that you have lied in your last conversation, according to a study from 2002.[1] Sometimes, we tell a lie to trick somebody by giving them false information. Other times, we just leave out a bit of information which makes our statements misleading. And sometimes our deceptions are quite innocent: we lie to be polite and to protect others from harm. At times we even lie to ourselves.

How Often Do We Lie?

The science of deception is a tricky one. How do you determine how many times the average adult lies? Some studies have asked this question in anonymized surveys. In a study from 2010, Kim

Serota and her colleagues from Michigan State University tried to design the research as carefully as possible. The 'lying survey', completed by 1,000 participants, was hidden between other online surveys (meals, cat products, and water softeners) to draw less attention to the subject of lying. The survey gave many examples of lies, including subtle and innocent ones, to make sure that people answered with the same definition in mind. Despite all these efforts, what the scientist relied on were honest self-reports. They found that, on average, people admitted to telling one to two lies a day. But how do we know that people did not lie about their lying? An intriguing result of this study is the exact distribution: three out of five people said they had not lied at all in the previous twenty-four hours. The remaining two out of five therefore account for the average one to two lies a day, meaning that some of these people lie a lot more than that. In fact, one subject reported to have told fifty-three lies in the previous twenty-four hours (the highest number was actually 134 lies, but this subject was treated as an outlier and not included in the analysis).[2] So does our society consist mostly of honest people who lie very little and some compulsive liars?

Let us come back to the study from 2002 that we mentioned in the beginning.[1] Robert Feldman and his colleagues from the University of Massachusetts invited pairs of psychology undergraduate students, who did not know each other, to participate. They were told that the study was about how people interact when they meet. Student A was told to appear likeable, to appear competent, or to act as they normally would when meeting someone new; student B always received instructions to act naturally. The following ten-minute interaction between the students was filmed without the participants being told about it (of course the material was only used if they later gave consent). Afterwards, student A watched the recording of the conversation and was asked to identify any lies he or she told. The researchers found that people lied on average about two times during the ten-minute interaction. But again, the distribution gave a more detailed picture: two out of five students said they did not lie at all; the remaining three out of five told on average about three lies. Overall, people lied significantly more when they tried to appear likeable and competent than in the control condition, where they were instructed to just get to know the other person. You could argue that in a real-life scenario you would not make an extra effort to appear likeable and competent because you have been told to do so by a scientist. The researchers might have encouraged lies in these conditions. The control condition might be the best comparison to everyday scenarios. But many of us are probably still motivated to be seen positively. Think about it: how often do you meet somebody and truly not care what they think about you? But this motivation is natural rather than artificial and instructed. So how many times did people lie when they had a casual chat with somebody they just met? Approximately once in ten minutes. Lying seems to be a part of many kinds of conversations. The researchers also looked at gender differences and found that men and women lied equally often.[1] But how well can a sample of psychology undergraduate students represent the general public?

There is evidence that the frequency of lying changes considerably through life. Evelyne Debey and her team from Ghent University recently tested over 1,000 visitors to an Amsterdam science museum aged between six and seventy-seven years. They reported how often they had lied in the past twenty-four hours, and afterwards their lying skills were tested. For this test, the participants answered questions such as 'Is fire warm?', or 'Are bananas red?'. Sometimes, they were instructed to deliberately give a false answer, sometimes to answer truthfully. The researchers then looked at the difference in reaction time between false and truthful answers. The faster people can tell a lie compared to telling the truth, the better their lying skills are. Also, the researchers calculated how accurately people performed the task: if they really did lie when instructed to do so. For example, when you are supposed to lie in response to the question, 'Are bananas red?' but you answer, 'No', then you would have a lower accuracy of lying.

So who are the best liars? Young adults, according to this study. They had the highest accuracy of telling lies when asked to do so, and they were also very quick. In contrast, young children and seniors lied less accurately. Older adults also took longer to tell a lie compared to telling the truth. This finding is related to cognitive control (more about this in Chapter 6) and indeed very young and elderly people showed poorer inhibitory control on a separate task. Telling a lie requires you to suppress the truth. The brain circuits needed for this inhibition do not fully mature until the mid twenties. Therefore, young children improve with age. However, the ability to self-control also declines later in life, which is why older adults struggle. To test this hypothesis it would be useful to follow up this experiment with a neuroimaging study and to relate lying skills directly to the corresponding brain circuits. But the behavioural results give us a good idea about lying frequency throughout life.

Now we have a better idea of who is very good at lying. But who lies the most? The self-reports from the study show a similar pattern to the lying skills and ability to self-control: young children and seniors lie the least, while adolescents and people in young to mid adulthood lie more.[3] Could it be that lying is more effortful for very young and very old people, and therefore they lie less often?

The biggest concern for studies relying on self-reports is honesty. How truthful are participants in identifying their own lies, given that this must be quite an uncomfortable and embarrassing experience?

How to Detect a Lie

If we do not want to rely on self-reports, what alternatives do we have? Can we ask people how many times they have been lied to by others? Would this give us a more honest and accurate result? The truth is that people are bad lie detectors. In an extensive analysis from 2006 including more than 24,000 participants, subjects were only 54 per cent accurate in detecting a lie. This is only a little better than a 50 per cent random guess of lie or no lie.[4]

People have been trying to detect lies for many centuries. The earliest methods were mostly based on superstition and faith rather than science. For example, certain hill tribes in Bengal made the accused lick a red-hot iron nine times. If the tongue was burned, the suspect was found guilty.[5] Later methods focused on physiological measures such as the pulse, suggesting that there was a growing insight into the body's response to telling a lie. A European nobleman from the Middle Ages is said to have convicted his wife of cheating by measuring her pulse. According to a report from an ancient book, his advisor put his hand on the wife's wrist while talking to her over dinner. When he mentioned the name of her alleged lover, her pulse went up. Unfortunately, she did not show the same reaction when her husband's name was mentioned and she later confessed the affair.[5]

The reliability of these early physiological methods was limited. Later research led to the development of the polygraph in the early 1900s. This instrument measures a range of physiological reactions: pulse, blood pressure, breathing rate, and skin conductance (influenced by the amount of sweat on your skin). You would think that such a rich data set could identify lies reliably, but studies show that the polygraph is only about 70 per cent accurate.[6]

The new kid on the block is fMRI. The number of studies using fMRI for lie detection has grown rapidly. There are at least two companies—'No Lie MRI' and 'Cephos'—which have commercialized the technique for lie detection. The method is advertised as 'the first and only direct measure of truth verification and lie detection in human history'.[7] The appeal is obvious: rather than measuring indirect body responses, the focus is on the organ that produces the lie: the brain.

In India, brain recordings have already been admitted as evidence in court cases (or proceedings). In the first case, in 2008, a court near Mumbai allowed recordings from electrodes on the scalp as evidence (a technique called electroencephalography or EEG). The suspect, Aditi Sharma, allegedly poisoned her ex-fiancé. The electrical recordings supposedly proved that she had detailed knowledge of the events. The brain recordings and circumstantial evidence eventually led to her conviction,[8] but she was later released based on the fact that the evidence was insufficient.[9] At the time of writing, fMRI-based lie detection tests have not been allowed in court to our knowledge. There have been several attempts in US courts to include fMRI scans, mostly involving the two companies mentioned earlier. However, these attempts have been unsuccessful so far. In 2009, defence attorneys tried to include evidence collected by No Lie MRI that supposedly proved the innocence of the suspect in a child sex abuse trial,[10] but the evidence was later withdrawn.[11] In 2010, Judge Tu M. Pham recommended that a report from Cephos involving fMRI lie detection should not be admitted as evidence in a fraud trial. Judge Pham concluded that the technique was not sufficiently reliable.[12] Let us have a look at the studies and see what brought the judge to this conclusion.

Brain Networks of Deception

One of the first studies that used fMRI to investigate deception was published in 2001 by Sean Spence and his group from the University of Sheffield. Ten male subjects answered thirty-six questions about their activities on that day, such as making their bed or taking a tablet. They answered the questions once, truthfully in the beginning. Then, while lying in the fMRI scanner, they answered each question twice more—once they were instructed to tell the truth, once to lie, by pressing a 'yes' or 'no' button. The researchers then compared the subjects' brain activity between the truth and lie conditions. While the participants were slower when answering a question untruthfully, they also showed greater activation in a part of the ventrolateral prefrontal cortex.[13] This brain area sits in the front part of the brain, halfway above the eye and ear. It is involved in many processes, one of which is response inhibition (although this is usually investigated in the context of motor inhibition).[14] Other studies link this part of the ventrolateral prefrontal cortex to language processing,[15,16] attributing intentions to others,[17] and decision-making.[18] The design of this study illustrates one point: the early studies were conducted to discover the 'brain network of lying'—to investigate the areas involved in producing a lie (or suppressing the truth). The researchers were not on a mission to prove the reliability of fMRI for lie detection. In fact, they did not create an algorithm to identify lies.

In 2002, Daniel Langleben and his colleagues from the University of Pennsylvania wanted to learn more about the brain regions involved in deception. They recruited eighteen subjects (males and females) from the university community. They were given a playing card and told to put it in their pocket. Then, while in the scanner, they were shown pictures of a few different playing cards and asked, 'Do you have this card?'. They were told to answer truthfully for all cards except the one in their pocket, but to deny the possession of the card they did have. Again, researchers found some brain regions that were more active during deception compared to telling the truth.[19] Among them was the anterior cingulate gyrus, which is a deep brain structure above the eyes towards the centre of the brain. It is involved in response inhibition and the detection of errors,[20] but also emotion regulation[21] and many other functions.

In 2008, Daniel Langleben led another study on lie detection. Participants picked a number between 3 and 8. The subjects were then shown numbers from 1 to 9 while lying in the scanner and told to deny having picked any of them. When they lied, the researchers saw a different activation pattern compared to telling the truth, including higher activation in the ventrolateral prefrontal cortex and anterior cingulate cortex, among other regions.[22] However, we should treat the results of this study with caution, until replicated by another laboratory. Although the article has been peer reviewed, the work was funded by a grant from No Lie MRI.

Accuracy of fMRI for Lie Detection

There is a growing literature on theories and mechanisms of deception, and it is certainly important and exciting to find out more about the brain networks involved in lying. However, if fMRI should ever be used in courts for lie detection, we will need larger studies that determine the levels of accuracy of the technique, especially at the single subject level. The majority of studies focus on mechanisms of deception and differences between truth and falsehood at the group level rather than the individual. However, to have a reliable technique for court, we will need to be able to separate truth from lies in every individual subject investigated.

Only a few studies have addressed the issue of accuracy so far. One of the first studies was published in 2005 by Daniel Langleben and his colleagues from the University of Pennsylvania. They used the playing-card paradigm explained in the previous section and tested twenty-six male undergraduates. First, they looked at differences between falsehood and truth conditions at the group level. They used a statistical analysis which is the standard in the field. With this strict analysis the team did not find any significant differences between lying and telling the truth, so they did not replicate their own findings from 2002. Interestingly though, the researchers used the data to generate a computer model. Such a model has to be tested with subjects who did not contribute to the model, which is what they did next. Four new male subjects were tested with the same paradigm, and the model then tried to classify instances in which they lied and those in which they responded truthfully. The model's predictions were 76.5 per cent accurate,[23] which is only a little better than the polygraph. Langleben's technology was later licensed by No Lie MRI, so there must have been some potential there. Indeed, the same data set was later reanalysed by the group with more sophisticated methods. Using machine learning techniques, the model achieved an accuracy of 88.6 per cent.[24]

In real-life court settings, the scenarios are likely to be more complex than denial of possessing a playing card. Andrew Kozel and his colleagues from the Medical University of South Carolina have designed a more realistic scenario. A mock crime was committed by sixty-one participants recruited from a university community. On the day of testing, they were instructed to 'steal' a watch or a ring from a drawer and to put the object into a locker together with their belongings. Afterwards, they had to answer four types of questions while undergoing fMRI: asking whether they took the ring, took the watch, neutral questions (such as 'Do you like the beach?'), and control questions about other immoral acts ('Have you ever cheated on a test?'). The participants denied having taken either object. To motivate them to lie convincingly, the subjects were told that they would receive an additional $50 if the experimenter could not tell when they were lying (although in actuality, they received the bonus no matter how they performed). Therefore, the subjects were highly motivated to beat the test. The model was built using the neuroimaging data from half of the subjects (thirty participants) and put to the test using the other half. Again, the neuroimaging analysis showed areas of greater activation during deception around the anterior cingulate cortex and several frontal areas. The accuracy of lie detection was 90 per cent, which is very impressive. The authors report that some participants even tried to use clever strategies, such as pretending they did not steal anything or imagining a specific place, 'none of which reduced the chance of having the lies correctly determined'.[25] You might not be surprised to hear that this technique has also been licensed—this time by Cephos, who also funded the study.

The highest reported accuracy of fMRI-based lie detection at the single subject level is 100 per cent, according to a study from 2011 led by Giorgio Ganis and his research team without conflicts of interest and with affiliations in the USA, UK, and Netherlands. Twelve undergraduates from Harvard University were told to lie about their date of birth. In the scanner, they were shown different dates and asked to answer the question 'Do you know this date?'. Subjects responded with 'no' even if they saw their birth date, with the exception of one 'target date', which subjects had to study before scanning. The participants were instructed to respond with 'yes' to this target date to ensure that they paid attention and did not just press 'no' repeatedly. The model for detection was trained with data from eleven participants and put to the test with data from the remaining participant. Strikingly, the model was 100 per cent accurate in distinguishing a lie from the truth. The authors explain this remarkable finding with the special meaning of a birth date, because it is autobiographical rather than artificially meaningful.

The really interesting part, however, is that the subjects were able to trick this seemingly perfect model. Participants were trained to use countermeasures: they were instructed to move the left index finger, the middle left finger, and the left big toe subtly for the majority of the irrelevant dates, just before pressing the button. They still responded with 'no', which was truthful, but it turned out that these subtle countermeasures disrupted the model. It classified lies and truths accurately in only about a third of the cases, which would be useless in a legal setting. The authors explain that these countermeasures give special salience to the irrelevant dates and could also be achieved 'by performing other mental actions (e.g. recalling a certain episode from memory)'.[26] In conclusion, if a suspect had enough knowledge of possible irrelevant stimuli that might be shown to him or her during scanning, even the most accurate computer model can be fooled. Think of poker players or actors, who are naturally skilled at lying—would they even need to be trained to trick the system? It is intriguing to think what we would find if we tried to replicate this experiment using 'skilled liars'.

The Limitations

Apart from skilled liars using clever strategies, what are the limitations of fMRI-based lie detection? A recent meta-analysis by Martha Farah from the University of Pennsylvania and her colleagues found that not a single brain region was consistently active during lying across all twenty-three included studies.[27] This finding gives a flavour of the inconsistencies of certain neuroimaging research. Task design, stimuli, characteristics of the scanner, analyses, experimenter, emotional state of the subject, and many more variables can have a considerable influence on results. To make the technique reliable in a legal context, we will need to know exactly how many errors occur and develop standardized protocols which produce robust findings that can be replicated.

Another important limitation is the condition of cooperation. The participants included in the studies participated for course credit or financial reimbursement. They were motivated to follow instructions, sometimes even by being promised a bonus if they performed very well. But if you were a suspect in a trial and the court ordered an fMRI-based lie detection protocol, you would have one simple way of making the data useless, without sophisticated countermeasures: you would just have to move your head during scanning. Even the slightest movements can make neuroimaging data useless. Researchers often have to exclude a considerable number of subjects due to too much head movement, even if the subjects did not move on purpose—this is how sensitive the technique is to movement.

You can also look at the problem of movement the other way around: even if people wanted to cooperate and undergo fMRI scanning, some are not capable of lying still in the scanner. For example, patients with Huntington's disease or some severe cases of ADHD would not be able to lie still without medication (and who knows what effects medication could have on brain networks involved in lying). Likewise, people with metal implants cannot go into an fMRI scanner (due to the strong magnet), nor those who suffer from claustrophobia (due to the tight tube). Would neurodegenerative diseases or brain injury alter the neural circuitry involved in lying?

And what about people with personality disorders? How does the brain of a psychopath look when he is telling a lie? So far, we have only discussed studies in healthy subjects, which is how the majority of studies on deception are conducted. In contrast, Weixiong Jiang and his colleagues from Central South University in China have investigated lying in thirty-two young offenders with antisocial personality disorder (according to the authors, the participants had 'committed misdemeanors').[28] They classified the subjects as non-liars, mild liars, and severe liars, according to the score based on an interview. Subjects were asked about their attitude towards lying, how successful their past lying was, how skilled they were at lying, and how often they lied. Although we cannot be sure that the subjects were honest, the scientists did find differences between the groups. When comparing differences in brain activation between answering truthfully and deceiving, the contrast was the strongest for the non-liars, less strong for the mild liars, and weakest for the severe liars. Although the researchers did not build a computer model to classify the responses, we can get the impression that deception would be harder to detect in skilled liars.[28]

What if the subject tried to fit the lie into a story and to memorize the scenario? Does this give rise to the same neural activation as recalling a true memory? Giorgio Ganis from Harvard University and his team addressed this question in 2003. Their ten participants answered questions about their most memorable holiday and work experience. Based on their answers, the scientists then prepared questions for the fMRI scanning session one week later. On the day of scanning, participants were instructed to generate an alternative, untrue scenario for either the vacation or work experience. The researcher helped them to create a coherent and consistent scenario based on untrue statements, such as spending the holiday in a different city or travelling by plane instead of by car. The subjects then rehearsed and memorized this false scenario. Afterwards, the participants went into the fMRI machine and answered three types of questions. For some questions, they were instructed to tell the truth, for some they would answer using details from the rehearsed scenario, and for others they had to spontaneously come up with a lie. The scientists found that the rehearsed lies gave rise to a different activation pattern from spontaneous lies. In addition, both showed a pattern that was significantly different from telling the truth.[29] This interesting study shows that lying is not one homogeneous act that activates the same brain regions consistently. Different types of lies have to be taken into account to develop reliable lie-detection techniques.

How realistic are the studies? First of all, subjects are usually instructed to lie in one condition and told to tell the truth in another one. There is very little room for personal choice—the subject making the free decision to deceive or to answer truthfully. Being told what to do by an external person may well have an influence on the brain activation we see. Moreover, most of the study participants were undergraduate students or people from a university community. It is unclear if lie detection works with similar accuracies in less educated, older, or very young people. In addition, the emotional state of the subject is very likely to have an influence on brain activation. A scientist or attorney might be convinced that brain scans prove that the suspect is lying, but the signal could also reflect extreme fear, disgust, or anger in response to a stimulus or to the overall situation. How would we control for this in a legal setting, where the defendant is very likely to be under increased pressure and stress?

Another issue that has to be addressed is the challenge of false memories. The fMRI technology looks very impressive, and expert witnesses can have a great appeal to the jury. But even the most advanced fMRI-based lie detector can only reflect the subject's beliefs. It is not an absolute and objective reflection of the truth, and has to be analysed critically. Imagine a subject who believes a false detail to be true. This could lead to false self-incrimination, for example by a suspect in a murder trial who believes they arrived at the crime scene at 2 a.m.—the time of the murder—when they actually arrived twenty minutes later. Due to the extreme stress of seeing the victim and being involved in the trial, the memory might become unreliable. Having heard the time 2 a.m. several times during interrogation, the suspect might come to believe this to be his or her own time of arrival. Neuroimaging shows increased activation to the stimulus '2 a.m.' and the prosecutor uses this information as evidence in court.

The jury is impressed by the neuroscientific evidence (in fact, studies have shown that this kind of evidence is very persuasive: non-experts rated bad explanations as more satisfying when neuroscientific evidence was included, even if that evidence was irrelevant in the context[30]). Surely, when a brain scan shows the time of arrival as 2 a.m., the jury should think this has to be the truth? By no means. All a brain scan can possibly reveal is the suspect's beliefs, not an absolute truth. This is an imaginary scenario for now, but if this kind of evidence was ever admissible in court, the jury would have to be educated about the limits of the technique.

The final point we would like to discuss is the issue of human rights. Defendants have the right to remain silent, to avoid self-incrimination. However, you could argue that our brain is never silent. Even if you just sit still or sleep, there is always activity in your brain. So when you are forced to undergo fMRI scanning for lie detection and you are asked a question, you might prefer to remain silent but your brain could still show a reaction that is meaningful in this context. If this reaction can be interpreted and used in court, do you lose the control of information you are unwilling to provide? Suspects can already be forced to provide fingerprints and blood/saliva samples for DNA testing. But one might argue that brain activation is a completely different category. Being coerced or forced to undergo this fMRI-based lie-detection test might eventually threaten the right of privacy of thought.

While the evidence is encouraging and advancements of science will certainly lead to the development of more reliable and accurate lie-detection techniques, there are several questions that we need to address. Questions of reliability in different types of participants with different skill levels of lying have to be addressed as well as the option of using countermeasures. Different types of lies will also need to be investigated and the limitations of the technique will have to be communicated clearly. We will need standardized protocols and minimal error rates, validated in large-scale studies which are conducted without conflicts of interest, to ensure the reliability of the technique. Finally, the dangers of violating human rights will have to be evaluated and put into an ethical framework before we can even think about widely admitting fMRI-based lie-detection evidence in court.

5

HOW MORAL IS YOUR BRAIN?

Let us do a thought experiment. A train carriage is racing down a hill. Its brakes do not work. It is approaching five workmen on the track. They will be killed if you do not act. The only alternative is to turn the trolley onto another track to the right, but there is also one workman there. What will you do? Kill one man to save five? Or not act at all?

This is an example of a paradigm that has been extensively studied. Originally, it was a philosophical thought experiment to explore morality.[1] More recently, the trolley scenario has also been used in neuroscientific studies to investigate cognitive processes and neural networks involved in moral decision-making.

Either decision—not acting at all and allowing five people to die, or turning the trolley and killing one person—violates a moral principle: not causing harm. It is a moral dilemma. What would most people do in a situation like this? Surveys show that the vast majority would switch the track of the trolley to save the five workmen, sacrificing the one workman on the other line.[2]

Let us look at a very similar problem. Again, there is a runaway trolley about to kill five people on the track. Only this time, you stand next to a large person on a footbridge over the tracks. You can only save the five people by pushing the stranger off the bridge, which will kill him but stop the trolley. What will you do this time? Strikingly, most people would not push the man off the bridge in this scenario.[2] This is very interesting, because from a very basic point of view, the two scenarios involve similar decisions:

(1) act and sacrifice one man to save five people, or
(2) do nothing and let five people die.

Yet, people decide very differently. Why might this be?

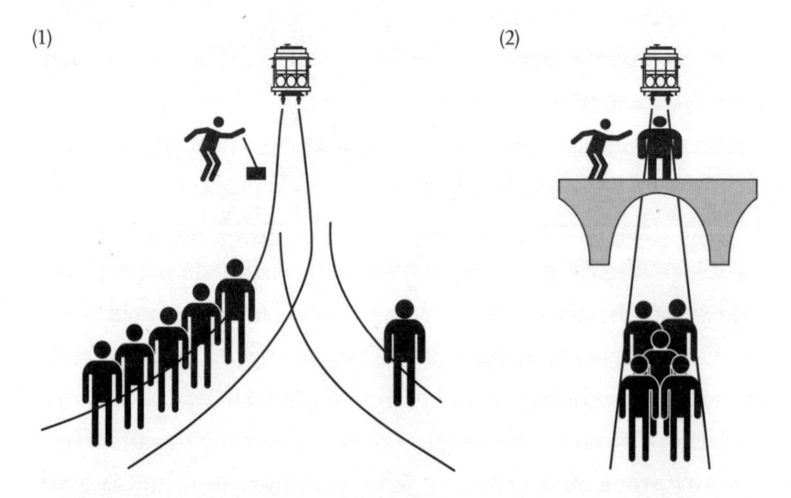

Fig. 4. Illustration of two types of trolley dilemmas. (1) Will you turn the trolley onto the track with one workman? (2) Will you push the large person off the footbridge to stop the train?

Different Types of Moral Decision

The philosopher and neuroscientist Joshua Greene conducted studies at Princeton University to investigate this question. He argued that the two scenarios have one important difference: the emotional involvement of the participant. The team compared two types of moral dilemmas given to participants while undergoing fMRI. One type of scenario was more personal, like actively pushing the man off the bridge; the other type was more impersonal, for example switching the track of the trolley. The researchers found significant differences in brain activation between the personal and impersonal moral scenarios. Most importantly, areas associated with emotion processing were more strongly activated in personal scenarios. One of them is the ventromedial prefrontal cortex,[3] an area just above the eyes. This research suggests that emotional processing plays a key role in moral decision-making, specifically for personal scenarios.

These findings were supported by studies in patients who had sustained damage in the ventromedial prefrontal cortex due to a burst blood vessel. These patients remain intellectually capable, but have impairments in emotional processing. For example, they lose interest in the personal lives of people close to them and ignore social rules. Elisa Ciaramelli and her colleagues from the University of Bologna showed that these lesion patients were more likely to choose pushing the man off the bridge in order to save the five people than participants without brain damage. In contrast, their responses to the impersonal moral dilemmas (switching the track) and also non-moral dilemmas (e.g. repairing an old television or buying a new one for the same price) were similar to the healthy participants'.[4] The ventromedial prefrontal cortex seems to play a crucial role in processing the emotional component of moral judgements.

All the moral scenarios we just discussed resulted in harm. People's decisions probably involved reasoning to reduce these severe damages. But how about situations where no harm is caused?

Morality of Intention

How do you feel about somebody who attempts murder, but fails—maybe because we catch this person in the act or because the victim had good self-defence skills? Is the intention itself amoral?

Most of us would make a judgement based on the outcome *and* the intention. This is also how our legal system works. The attempt to cause harm can be punished, even if nothing happened in the end, such as attempted murder. This kind of reasoning requires skills to understand someone's intentions, to put yourself in another person's shoes. This process is called 'theory of mind' and it is a key component of moral judgement.

Most of us can understand the difference between accidental harm and intended harm. We would be more willing to forgive somebody who accidentally splashed water on our laptop than we would an evil competitor who attempts to destroy our computer but is caught in the act. In one case, actual harm was done; in the other one, your laptop remains untouched and your data are safe. And yet, we are harsher with the evil person than the clumsy one. This general ability to put emphasis on intentions is common to most of us, but there are individual variations. Some people might still want to put blame on your clumsy colleague, even though he or she had no evil intentions.

Liane Young and Rebecca Saxe from the Massachusetts Institute of Technology have looked at these scenarios combined with neuroimaging. They found that the tendency to diminish blame for accidental harm was correlated with the activity in the right temporo-parietal junction.[5] This brain region is located towards the back of your brain and is one of the key areas for reading other people's intentions.[6] The higher the activity in this brain region, the lower the assignment of blame to the person causing accidental harm.[5] It seems that this region plays a role in understanding the innocent intentions of the agent. Networks in the brain that are involved in social interaction are key for making moral judgements. What other regions are involved in moral thought and behaviour?

In Search of the Morality Centre

A range of researchers have set out to conduct fMRI studies to find the neural basis of morality. One of the first experiments of this kind was conducted by Jorge Moll based at the LABS & Rede D'Or Hospitais in Rio de Janeiro. In 2001, Moll and colleagues invited ten people to listen to statements and to judge them as either 'right or wrong'. Some of these scenarios were moral ('They hung an innocent'), while others were statements without a moral component ('Stones are made of water'). The researchers collected fMRI data while subjects were making judgements.[7]

In another experiment, Jorge Moll and his team presented seven participants with social violations. These were either of a moral nature ('The elderly are useless') or non-moral ('He licked the dirty toilet'). Subjects then had to judge whether the statements were right or wrong.[8] The same group took the paradigm one step further in 2002. Their participants viewed images showing unpleasant scenarios. Some of the pictures showed moral violations, such as physical assault, while others were unpleasant but without moral content, such as body lesions.[9] If you think that these experiments sound distressing, you are probably right. But bear in mind that these are volunteers who had the chance to stop the experiment at any point.

In all these studies, the researchers were interested in the same analysis: which brain regions are active when thinking about moral but not non-moral contents? In other words, where is the 'morality network'? One region that was activated during moral scenarios was the ventromedial prefrontal cortex, the association that was observed by Joshua Greene. Another common finding of these studies was activation in the superior temporal sulcus, which is located above the ear and stretches along the side of the brain. This brain region has been shown to be involved in the perception of speech, motion, faces, and in the theory of mind, to name only a few processes; you can probably see why it has been called the 'chameleon of the human brain'.[10] This pattern is true for all the 'morality' regions found in these studies: the researchers did not find a single brain area of morality that is not also active during other processes.

In seeking a key brain area for morality, scientists actually found networks involved mainly in emotional and social cognition. This suggests that morality is a complex process that involves many different aspects of cognition. Rather than a centre of morality, moral decision-making appears to rely on many parallel, cooperating systems and brain areas. If we wanted to identify such a moral brain area, this region would need to be activated by all moral scenarios—not just emotional ones or where you are specifically thinking about intent. In addition, such a region should not be activated in response to any other cognitive process, such as thinking about your colleague's mood outside a moral context. But even if we cannot pin morality down to one centre, we can still get a pretty good idea of the network necessary for it thanks to fMRI studies.

There seems to be a brain signature for viewing moral scenarios and making moral judgements. Even though there are individual differences between participants, the general pattern is similar for most people. But are there people who do not show the same activation?

Abnormal Morality

Now that we know more about the typical network of morality, we can think more about deviations from the norm. What kinds of individuals might show a different brain activation in response to moral content? We have already seen that patients with lesions in the ventromedial prefrontal cortex respond differently to the trolley dilemma (they were more willing to push the large stranger off the bridge to save the five people on the track). Are there any other conditions with similar abnormalities?

One personality disorder is obviously linked with alterations in moral behaviour: psychopathy. Psychopaths often disregard moral values of our society, manipulate, cheat, steal, or even kill, but show no remorse.[11] However, there are many misconceptions linked to the disorder. Contrary to popular belief, psychopaths are able to experience empathy. They are able to understand another person's feelings when asked to do so. However, their 'spontaneous' empathy, that is their level of empathy without instructions or special focus on it, is very low compared to healthy people.[12] We have seen before that thinking about another person's feelings is a core process of morality. We know that psychopaths often disregard moral rules, but do they understand them?

Several studies have shown that psychopaths can tell right from wrong. They are able to judge that a moral transgression, such as a child hitting another child, is not permissible. Interestingly though, psychopaths seem to be unable to distinguish between moral and conventional transgressions. In other words, they do not seem to make a difference between a violent child and one that is talking in class—both transgressions are rated as serious and not permissible. Healthy people, in contrast, rate the moral transgression of violence as more serious and less permissible.[13,14] These findings suggest that psychopaths have deficits in moral processing. Their behaviour is abnormal, but what about their brain activity?

Carla Harenski and her team went to a prison in North America to find out. They scanned the brain activity of sixteen psychopaths and sixteen non-psychopathic offenders while they viewed pictures. Similar to previous studies, the pictures were unpleasant. Some contained a moral violation (e.g. somebody breaking into a house), while others did not (such as a mutilated hand). The criminals had to judge if the picture showed a moral transgression, and if so, how severe it was. The ratings were very similar in psychopaths and non-psychopaths. Both groups were able to identify moral transgressions and rated them similarly severe. One could expect that moral processing is therefore similar in the two groups. However, the brain scans told a different story. Non-psychopathic offenders showed different brain activation when viewing moral vs non-moral pictures. Specifically for moral pictures, they showed a greater activation in the ventromedial prefrontal cortex compared to non-moral pictures. Psychopaths showed no difference between the two conditions. Their ventromedial prefrontal cortex was equally active when viewing moral and non-moral pictures.[15] Collectively, these findings suggest that healthy people view moral scenarios as 'special' or more salient. Psychopaths just do not seem to have that special category reserved for morality.

How should we use this information? Could brain scans be used to diagnose psychopathy in the future? And if people show this abnormal brain activation during moral judgement, does this automatically mean they are all psychopaths? Moreover, if we had a clear diagnostic tool of psychopathy, indicating that the people in question are likely to disregard our moral values, should we incarcerate them before they can cause any harm? Would that be the responsible thing to do?

It is too early to say if brain scans will ever be precise and specific enough to use them for a clear diagnosis. At the moment, the studies are too small and effects are too weak. The research examines group differences and not individual differences. Furthermore, there is an important point which is easily forgotten: even though brain scans give us a breathtaking wealth of information, they cannot predict the future. Not every psychopath will become a criminal: only the risk is likely to be higher. Evidence indicates that psychopathy affects less than 1 per cent of the general population,[16] but between 3 and 23 per cent of prison populations,[17–20] depending on study methodology and country. But these numbers do not tell us which psychopath will become a criminal and which one will not. Incarcerating individuals before they have any intention of committing a crime is only material for science fiction. Before we can justify such drastic measures, we will need major breakthroughs in science and technology. Our current fMRI systems are certainly not a crystal ball. Who knows if we will ever find one?

Abnormal Morality as an Excuse?

It is widely accepted that fMRI scans cannot infallibly predict the future. But can they explain or even excuse something that happened in the past? Could they be used as evidence in the courtroom to reduce the sentencing of a defendant?

The year 2009 saw a small revolution of the legal system: for the first time, fMRI scans were admitted as evidence in court. The defendant was Brian Dugan, a 52-year-old man from Illinios, who was already serving a pair of life sentences for the rape and gruesome murder of a 27-year-old nurse and 7-year-old girl.

Now he was on trial for raping and beating a 10-year-old girl to death. The prosecution sought the death penalty for Dugan, but his lawyers wanted to reduce the sentence by convincing the jury that Dugan was a psychopath. They hoped that this would reduce his criminal responsibility and spare him the execution. The lawyers approached neuroscientist Kent Kiehl, who had done studies on psychopathy for many years. He agreed to scan Dugan's brain, interview him, and to testify in court. His testimony and description of fMRI results were allowed in the trial, but not the scans themselves. After finishing his data collection and analysis, Kiehl took the witness stand and described the evidence supporting a diagnosis of psychopathy. However, the prosecution called another expert witness: psychiatrist Jonathan Brodie. He explained that the scan from 2009 could not possibly determine Dugan's state of mind back in the 1980s, when he committed the crimes. He also pointed out that most fMRI studies on psychopathy focus on group differences—how the *average* psychopath's brain differs from that of the *average* healthy person, and that the two groups would probably overlap and therefore we cannot make judgements about individuals. The studies about accuracy in individuals had simply not been done (and we still lack convincing evidence). The jury eventually voted to execute Dugan[21] (although the death penalty was abolished in Illinois in 2011 and Dugan is now facing a life in prison[22]). This judgment perhaps reflects how fMRI evidence in court still requires the determination of sensitivity and specificity at the level of individuals.

Admitting brain scans as evidence in court to explain an individual's behaviour opens a whole field of ethical problems. Defendants can be found 'not guilty by reason of insanity', indicating that they did not have the full mental capacity to make decisions and therefore cannot be blamed. In Chapter 6 we will meet a married schoolteacher who suddenly became a paedophile, but the discovery and removal of a brain tumour stopped the behaviour. Shortly afterwards, the immoral behaviour restarted, and it was found that the tumour had grown back. After its removal the man returned to normal again. In cases like these we are inclined to think that the individual should not be blamed, since the behaviour seems to be very tightly linked to a biological abnormality. We do not interpret the tumour as part of the man's personality, but rather as an abnormal influence that can be removed.

But how do we feel about psychopaths? Are they criminally responsible? If someone acts within a different moral system, does that make them less guilty, less responsible? When talking about criminal responsibility, it is important to note that we *are* our brains, since they give rise to our consciousness, our personality, our sense of self. All the decisions you make are down to neural activity, which results from your genes, your previous experiences, your environment. Saying 'My brain made me do it!' makes about as much sense as saying that J. K. Rowling convinced the author of the Harry Potter series to write seven books about the young wizard.

Despite this realization, in the case of the schoolteacher with a tumour, it feels reasonable to say that he is a victim of his tumour. Because both incidents happened at the same time repeatedly, we are inclined to think that one caused the other. But correlation does not necessarily mean causation. Just because two things happen at the same time or in the same direction, this does not mean they are causally related. However, the fact that the criminal behaviour only began after the tumour grew, and was extinguished when the tumour was removed (and that this happened twice) strongly suggests a causal link. The schoolteacher had never before shown behaviour of that kind—there were no signs of unusual preferences or illegal behaviour. The fact that the paedophilia was out of character and also started so suddenly might make us think that there was a strong, unusual influence.

But how is a tumour any different from, say, a genetic condition that causes your neurons to fire differently and changes your ability to make moral decisions? Is that your fault, if you have always been that way? Let us take this one step further. Suppose an individual has grown up with deeply amoral parents and he was taught that none of the societal rules of morality are valid. He fully endorses this thinking. When somebody crosses him, he thinks it is acceptable and within his rights to take that person's life (this is not meant to be a realistic description of any specific condition or case). Would you consider him fully responsible for his actions, even though he did not choose his upbringing? The current legal system allows the 'not guilty by reason of insanity' judgment, which usually requires the defendant to be in a state in which rational decisions cannot be made. However, the person amoral by upbringing may have been fully rational and have justified the actions to himself. When sentencing, the law does allow for mitigating circumstances. Where does responsibility end? When is it valid to excuse actions?

Using genes, the environment, or both as explanations or excuses for unfavourable behaviour is a slippery slope. How many genes might we discover that affect important brain circuits? How many adverse events in childhood, bad influences, and traumatic events could a skilled lawyer find for any of us? Does that really excuse behaviour? We have to draw the line somewhere. There are certainly cases where people suffer from severe mental illnesses, possibly causing delusions or hallucinations that make the person incapable of making rational decisions. But apart from these clear extreme cases, many of us would not want to make substantial adjustments to what is law. What kind of evidence is strong and convincing? It is up to society and the legal system to determine the constraints of criminal responsibility.

Altered Morality

We have looked at abnormal moral processing and possible implications for guilt in the legal system. But what about healthy people? If a person shows normal moral processing, can it be manipulated? Let us explore three ways of altering moral judgements: targeted magnetic fields, hypnosis, and drugs.

Thinking about intentions is a key component of moral judgements. This process is tightly linked to the right temporo-parietal junction, that area towards the back of your brain. The more active this region during moral judgement, the more likely a person is to excuse accidental harms, as shown by Liane Young and Rebecca Saxe.[5] The researchers were now interested to see if they could change moral judgements by manipulating this region with magnetic fields.

They invited twenty people to participate in the study. All of them first underwent an fMRI scan to locate the right temporo-parietal junction, because its exact location in the brain differs a little between people. When they had found the region, the experimenters applied transcranial magnetic stimulation to the subjects' scalps. In this technique, a coil which generates a magnetic field is placed on the desired position on the head. There are different settings to adjust how deep and how strong the stimulation is, to specifically target a certain brain area. The magnetic field passes through the skull and produces electrical currents in the brain. These currents then affect the neural activity in a certain brain area. Depending on the settings and the brain area, these currents can have different effects. Currently, the technique is trialled as treatment for a range of medical conditions, including types of depression,[23] stroke,[24] and schizophrenia.[25] Plate 3 shows a participant undergoing transcranial magnetic stimulation.

In this study, the researchers used transcranial magnetic stimulation to disrupt the activity in the brain area. They either stimulated the right temporo-parietal junction (the area of interest) or a control region on the other side of the brain that is known to have no direct involvement in moral processing. Subjects had to make moral judgements for different scenarios, with different outcomes and different intentions of the agent. Most importantly, some of the scenarios included agents with bad intentions, but who failed to cause harm. For example, one of these agents was Lauren who was camping in the woods with an acquaintance. They spotted some wild mushrooms and Lauren believed that they were toxic, causing painful convulsions and death. She offered the mushrooms to her acquaintance, who ate them and was fine afterwards. Was Lauren blameworthy? When the subjects received stimulation to the control area, they rated Lauren's decision to offer the mushrooms (and similar scenarios) as morally forbidden—a likely moral judgement given Lauren's bad intentions.

Strikingly though, when the researchers stimulated the right temporo-parietal junction, the subjects thought of the scenario as more morally permissible—Lauren was not judged as harshly any more. Some of the subjects received the stimulation immediately before making moral judgements, some during the decisions. In both groups, the disrupted activity of the right temporo-parietal junction made judgements of attempted harm less harsh—the intention seemed to be less important.[26] This is only one example of how moral decisions can be manipulated in healthy people. The technique had a powerful effect and applying it to other brain areas or with different settings may have even more dramatic consequences. It would be impossible, however, to use this technique incognito. You would certainly notice if somebody held a coil to your head passing a magnetic field through your skull and altering your brain activity. Are there more subtle options that can be applied without the subject knowing about it?

Thalia Wheatley and Jonathan Haidt from the National Institutes of Health and the University of Virginia conducted an interesting experiment using hypnosis. They brought sixty-four participants to the hypnotic state and induced disgust linked to a specific word (something innocent such as 'often'). Afterwards, the participants had no memory of what happened. They were asked to make judgements about different scenarios. Some descriptions included the disgust-inducing word and, strikingly, the moral judgements were more severe than for scenarios without the word. Importantly, this was not only the case for scenarios that most of us would rate as morally wrong, such as shoplifting or bribery. The disgust reaction even influenced scenarios involving no bad intentions—such as a student council representative who chooses topics for discussion that are appealing. Participants who read this scenario about appealing discussions with innocent intentions rated it more morally wrong when the description included their disgust-related word. In other words, context or environment, such as disgust words, can negatively alter moral appraisal. Some participants even tried to justify their ratings afterwards. One subject said, 'It just seems like he's up to something'.[27] Clearly, morality is a very complex process that somewhat relies on unconscious intentions and beliefs. Surprisingly, a single session of hypnosis was sufficient to show strong effects on moral judgement. Altering people's morality can be a powerful tool for manipulation. In the wrong hands, this could become a threat to society.

Hypnosis, however, only works on a small proportion of people, and the subject has to be open to the experience. This does not sound like an option to take over the world. But there seems to be a more powerful way to alter morality: drugs. A recent study led by Molly Crockett from the University of Oxford and University College London explored the potential of two drugs to manipulate morality. They used a very different paradigm to that of most previously published studies. The conventional moral dilemmas often describe scenarios of severe harm, such as death. Deciding whether you want to push a large stranger off the bridge might not be the best representation of realistic moral judgements that most people usually have to make. Most of us would (luckily) never face such a situation. In contrast, Molly Crockett and her team used a paradigm with very real harm—electric shocks. The participants received shocks with increasing intensity to establish their pain threshold and to get a real-life experience of the stimulus.

Afterwards, the subjects were asked how much money they wanted to pay to avoid a shock being given to them or to a person next door. The research team had previously shown that people were willing to pay on average 10p more to avoid shocking somebody else compared to themselves.[28] Causing harm to somebody else violates a moral principle and people seem to be eager to avoid this. The group was now interested to see if this moral behaviour could be altered by drugs. One of the medications they chose was citalopram, an antidepressant that affects levels of the neurotransmitter serotonin in the brain. Citalopram strongly increased harm aversion, both to self and others. Subjects were now willing to pay almost twice as much to avoid shocks, both for themselves and others. The drug did not change the perception of the shocks, but the willingness to avoid harm. In contrast, the drug levodopa, which affects dopamine levels and is commonly used to treat Parkinson's disease, showed a different effect. Participants on levodopa did not pay more to avoid shocks to others compared to themselves. Instead, they paid a similar amount for both conditions. The strong desire to avoid shocking somebody else seemed to be reduced by levodopa.[29]

These findings need to be replicated before we can draw any conclusions. It is very important to note that we should not expect altered morality in all people who take these drugs as treatments for medical conditions. For example, depressed patients would be expected to have very different baseline levels of serotonin, so the effects of citalopram are not comparable to the healthy subjects in the study. The same is true for dopamine levels in patients with Parkinson's, if they receive the drug with the optimal dosage. However, there are reports of Parkinson's patients who develop hypersexual behaviour or excessive gambling after taking their medication at excessive doses, known as dopamine dysregulation syndrome.[30]

This study suggests that just a single dose of a common drug can alter moral judgement. In the case of citalopram, this effect could be very desirable. Imagine a world where everybody is keen to avoid harm, even more than we are already. What differences could we expect for crime rates? What would be the consequences for war environments? However, changing people's beliefs and judgements long term with drug treatments carries the risk of side effects. Also, would we want our society to be pharmacologically enhanced, even if the effects seem positive? Or would we regard other forms of enhancement, such as education, to be preferable to drugs?

Risky Potentials and Potential Risks

There are many controversial issues surrounding morality. While early imaging studies wanted to find brain networks of morality, researchers now try to find abnormalities within the relevant neural networks in different patient groups. These findings could possibly be used to explain atypical moral judgements and behaviour. Evidence from fMRI scans has already found its way into the legal system, though only with limited success so far. Given the steady improvement of technology, it might only be a question of 'when', not 'if' fMRI will be used routinely in court. We will need new frameworks and rules to regulate such use.

Possibly of even more concern are recent studies suggesting that normal morality could be altered by relatively simple means. We explored magnetic stimulation, hypnosis, and drugs in this chapter, but there might be other, more effective ways of altering an individual's moral system. Would we want to use these findings to morally enhance our society? To improve moral behaviour in healthy individuals? Or to 'normalize' morality in those who show abnormalities? This might sound like a nightmare scenario to some and a utopia to others. How will society adapt to the intersection between neuroscience, morality, criminality, and the law?

6

ARE YOU IN CONTROL?

Stanford, California, 1960s. A 4-year-old child sits in front of three mini-marshmallows. The child wants to eat the sweet treats right away, but there is a catch. She can eat one marshmallow immediately or wait a while before she gets permission to eat two marshmallows.[1] This waiting period can be up to fifteen minutes, and many children struggle to wait. Some give up after less than a minute, some manage to resist the temptation a little longer, and some children successfully wait the full fifteen minutes. It turns out that this famous 'Stanford Marshmallow Experiment' might tell us a lot more than just whether or not pre-schoolers are willing to wait to get two marshmallows. The experiment was repeated with some variations in more than 600 children between 3 and 5 years old. This cohort has since been followed up regularly and tested for a range of measures. The children who were able to wait longer became more academically and socially competent adolescents, who were better able to cope with stress.[2,3] They also had a lower body mass index thirty years later.[4]

The study does not prove definitely that self-control at the age of 4 is an inherent quality that determines later success in life. Nevertheless, this landmark study by Walter Mischel and colleagues from Stanford University does highlight how genes, environment, and developmental factors interact to influence cognitive control. The study was followed up in 2013 by Celeste Kidd and her colleagues from the University of Rochester, with one important difference: the children experienced the marshmallow experiment in a reliable or unreliable environment. The researchers provided the children with some art materials, such as well-used crayons. They told the children that they could either use them now or wait for the experimenter to come back with better supplies, such as a new set of crayons. The experiment was manipulated so that all children had to wait, because the old crayons were placed in a tightly sealed jar which was difficult to open. In the reliable condition, the researcher came back with an exciting collection of brand new art supplies. In the unreliable case, the experimenter came back empty handed and said there were no other supplies after all. After drawing for two minutes, both groups then completed the marshmallow task. Children who had experienced the reliable environment waited on average four times longer than the ones who had previously been disappointed.[5] The study suggests that self-control is not the only factor that we have to consider in the marshmallow experiment. It might also be important whether or not the child has been previously exposed to an unreliable or unpredictable environment, or whether the child trusts the experimenter. If you think that you will not receive a larger reward anyway, it might be the rational decision to go for the smaller but immediate reward. There are some more limitations of the Stanford Marshmallow Experiment, for example the fact that some of the children might have been hungrier than others, which was not measured. Also, they were mostly children of academics, so they shared the same background (though similar results have been found in a sample from low-income households from the Bronx, New York[6]). And the follow-up studies were conducted on a smaller subset of participants (typically around 100 of the original 600+), because some could not be contacted or did not want to participate again.

The most important point to consider is that correlation is not causation. In other words, a third factor, such as unpredictability at home, could be important. The results of the marshmallow task are striking and they provide a nice narrative. This is often picked up by the media in an oversimplified way. Nonetheless, the link between waiting for two marshmallows early in life and being good in school, socially adapted, and skinnier is probably not that simple. Even the strongest correlations cannot explain the full variation in the data set, and not every participant follows the same trend. There might be people who give up after ten seconds but still perform very well in school. And there are people who waited patiently at the age of 4, but were overweight in their thirties. There are countless factors that contribute to all these measures: life is too complex to draw simple conclusions. The results of the marshmallow task do not give us a rule set in stone, but they may give us an insight into when self-control develops. If we could help children gain mastery over their environment and develop control over impulsive responses, we might be able to promote resilience and improve life and health outcomes.

Control Centres in the Brain

Despite these limitations, self-control is an interesting and important ability. Eating chocolate for breakfast, lunch, and dinner will likely make you obese and diabetic. Speaking your mind all the time will make your friends disappear. And who is going to employ you if you stay in bed and watch cartoons all day? To regulate our behaviours, we have developed mechanisms of control and inhibition.

Scientifically, cognitive control can be investigated in several different ways. One commonly used paradigm is the Stroop test, named after John Ridley Stroop who described it in 1935.[7] In this test, the name of a colour (e.g. green) is printed in an ink either matching the colour name (e.g. green ink) or not matching it (e.g. red ink). As a subject you would be asked to name the colour of the ink. Stroop discovered that subjects respond faster and more accurately when the ink matches the colour name. If not (e.g. the word 'green' printed in red ink), subjects take longer to respond and sometimes make mistakes (they say 'green' even though the ink is red). Most adults are better and faster at reading words than naming colours. When these two processes compete, the subject has to override the automatic response (reading the word) with a more effortful process (naming the colour of the ink) (see Plate 4).

In order to find the brain areas involved in performing the Stroop test and how they develop with age, Rachel Marsh and her colleagues from Columbia University invited seventy healthy participants between 7 and 57 years to perform the Stroop task while undergoing fMRI. In this developmental study, the youngest participants performed poorly, but performance improved in adolescents and was stable in adulthood.

The researchers then looked at activated brain regions and correlated their findings with age and performance on the task. Two important regions were discovered in this analysis: the right ventrolateral prefrontal cortex and the right lenticular nucleus.[8] The former is located behind your right temple, and the latter is part of a deep brain structure in the centre of your brain called the striatum. More and more research is showing that the connections between prefrontal cortex and striatum, called frontostriatal circuits, are important for a range of cognitive functions and their control, such as working memory, decision-making, and planning.[9] The more active these frontostriatal regions were in this study, the better the performance of the subjects was and the older they tended to be. A stronger engagement of this important network led to better cognitive control in the Stroop task.

How else can we measure self-control? For example, we can look at motor control, often assessed with versions of a go/no-go task. The name gives it away: participants are asked to respond in one (go) condition, but to withhold a response in another (no-go) condition. Seiki Konishi and his colleagues from the University of Tokyo used a very simple go/no-go task to conduct the first neuroimaging study in humans on this task in 1998. They scanned only five subjects who were asked to quickly push a button when they saw a green square (go) but not when they saw a red square (no-go). The researchers looked for brain regions that were more active during no-go than go trials—areas involved in response inhibition. They found a significant difference in the right inferior frontal sulcus,[10] a pronounced groove in the prefrontal cortex. Since then, many more imaging studies have been conducted. A study led by Adam Aron from our own laboratory examined response inhibition in patients with lesions in the frontal cortex and confirmed the importance of the right inferior frontal gyrus—the region under the right inferior frontal sulcus.[11] A later meta-analysis by Derek Nee and colleagues from the University of Michigan and Columbia University revealed that the right dorsolateral prefrontal cortex, located roughly above the facial hairline towards the side of the brain, showed the most prominent consistency across the go/no-go studies.[12]

Another way of measuring self-control is one we already discussed: delaying an immediate reward in favour of a later, bigger reward. However, since marshmallows are not the most attractive reward for most adults, neuroimaging studies with these paradigms usually offer money rather than sweets. Samuel McClure from Princeton University led a study on immediate and delayed rewards in 2004. The article, published in the journal *Science,* starts with a description of Aesop's old fable about the ant and the grasshopper. The grasshopper enjoys a warm summer day, chirping and singing without worrying about the future. Meanwhile, the ant works tirelessly to store food for the upcoming winter. The fable ends with the grasshopper dying of hunger when winter breaks in, while the ant can feed itself from the stored food. Our real-life scenarios are usually not quite that dramatic, but the authors remark that we 'seem to be torn between an impulse to act like the indulgent grasshopper and an awareness that the patient ant often gets ahead in the long run'.[13] In the experiment, fourteen Princeton students were faced with the decision between rewards. One was available early (immediately, in two weeks, in one month) while the other reward was available later (two weeks or one month after the early one). The early reward was 1–50 per cent smaller than the late one and ranged from $5 to $40. For example, the subject would choose between $20 today or $30 in two weeks, while undergoing fMRI. The researchers found that when the subjects chose immediately available rewards, they had a higher activation in parts of the limbic system. This network has been shown to be involved in the expectation of reward.[14] In contrast, when subjects chose the delayed outcome, they had a greater activation in regions of the lateral prefrontal cortex (in the front towards the side of the brain) and the posterior parietal cortex (towards the top of the brain). This activation probably reflects the involvement of higher-level cognition, including inhibitory control.[13]

Mario Beauregard from the University of Montreal and his colleagues took a different approach to self-control. They investigated the ability to inhibit sexual arousal. The team invited non-criminal male volunteers to watch erotic films while lying in an fMRI scanner. In one condition, they could enjoy the films and show their normal responses to it—they were allowed to become sexually aroused. In the other condition, the men were asked to suppress their sexual arousal by trying to distance themselves from the scenario. They had to become a 'detached observer'.[15] This inhibition led to activation of the right dorsolateral prefrontal cortex (behind the forehead) and right anterior cingulate cortex (a deep structure towards the front of the brain). Some of the other studies mentioned here have shown a similar activation of the dorsolateral prefrontal cortex in conditions of self-control, highlighting the importance of this region for control processes. The anterior cingulate cortex has been shown to be active in the evaluation of emotional information[16] and error-monitoring.[17] Importantly, none of the regions associated with sexual arousal were activated when the subjects suppressed arousal.[15] This shows that mental inhibition can have dramatic effects on brain activation. It is likely that the frontal cortex was exhibiting inhibition over areas of the emotional brain, such as the amygdala.

Just from these few studies we can see that the brain networks involved in cognitive control are not part of one unitary network. Derek Nee's meta-analysis mentioned earlier in this chapter compared over forty neuroimaging studies of cognitive control tasks, including the Stroop task, go/no-go, and others. The analysis revealed that each task was associated with a different brain signature and that only a few regions were activated by all tasks.[12] Moreover, it has been known for a while that if you perform well on one task of cognitive control you do not necessarily do well on another one.[18] It seems that cognitive and response control requires many different processes, including sustained attention, memory, and motivation. But even if we cannot describe self-control as a simple, unified concept, we can still find interesting differences between intact and impaired control.

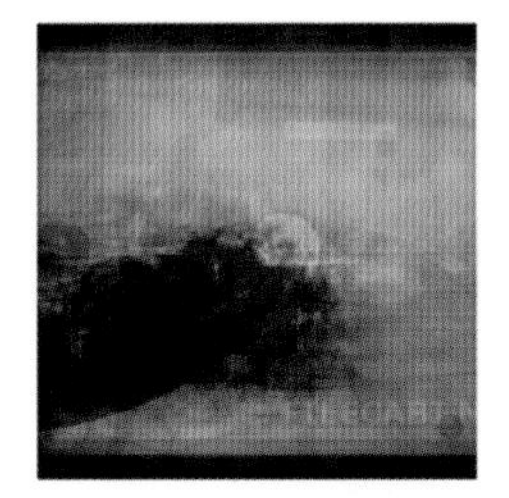
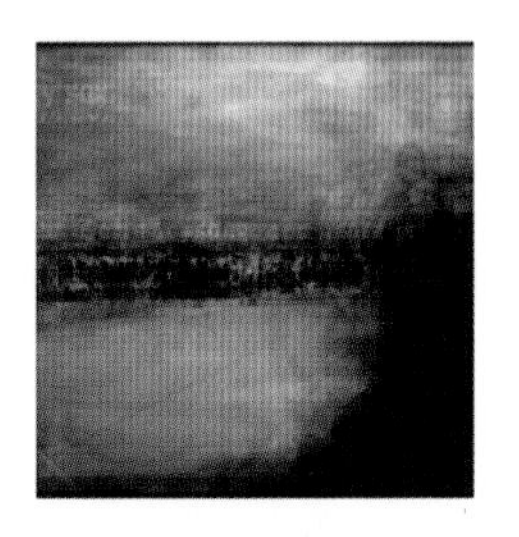

1. Segments of Hollywood film trailers reconstructed from brain activity that was measured using fMRI. The original trailers showed Steve Martin, an elephant in the desert, and an aeroplane.

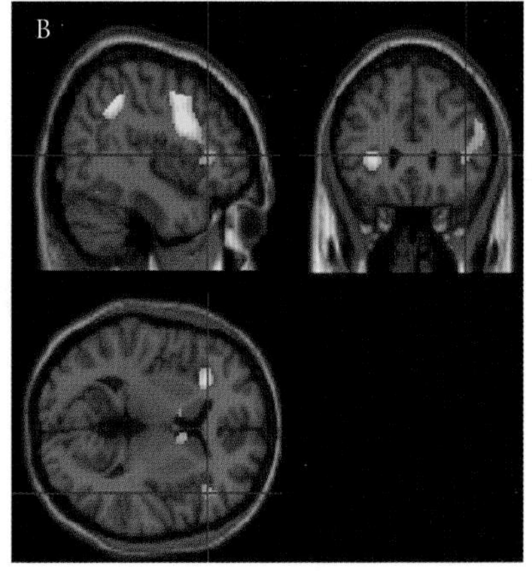

2. Adrian Owen—as we know him and an fMRI scan of his brain at work.

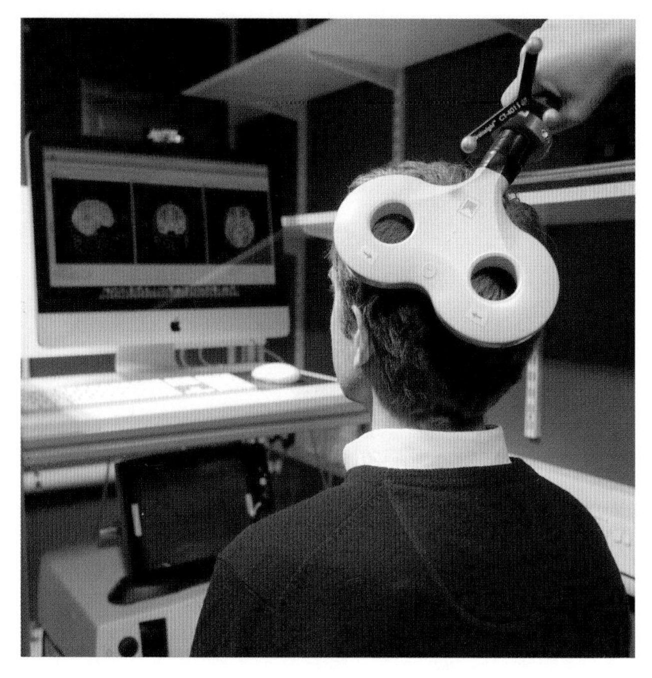

3. A participant in a transcranial magnetic stimulation experiment.

Easy:	**GREEN**	**YELLOW**	**RED**	**BLUE**
Hard:	**GREEN**	**YELLOW**	**RED**	**BLUE**

4. Illustration of the Stroop effect: Name the colour of the ink, not the word!

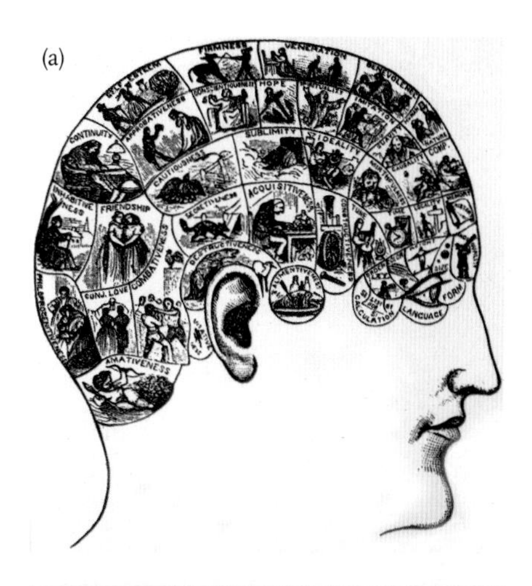

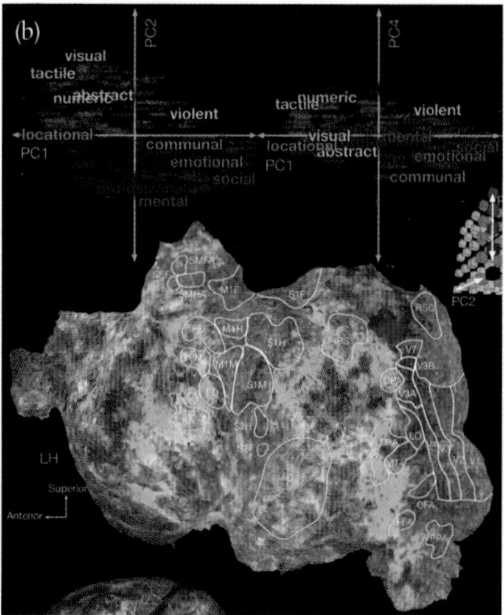

5. A phrenological chart, showing the alleged location of certain personality traits (a), vs the more visionary approach of mapping the semantic atlas of the human cerebral cortex (b). Image (b) represents the semantic map of one subject: green areas are mostly related to physical and perceptual categories, red and purple areas represent human-related categories.

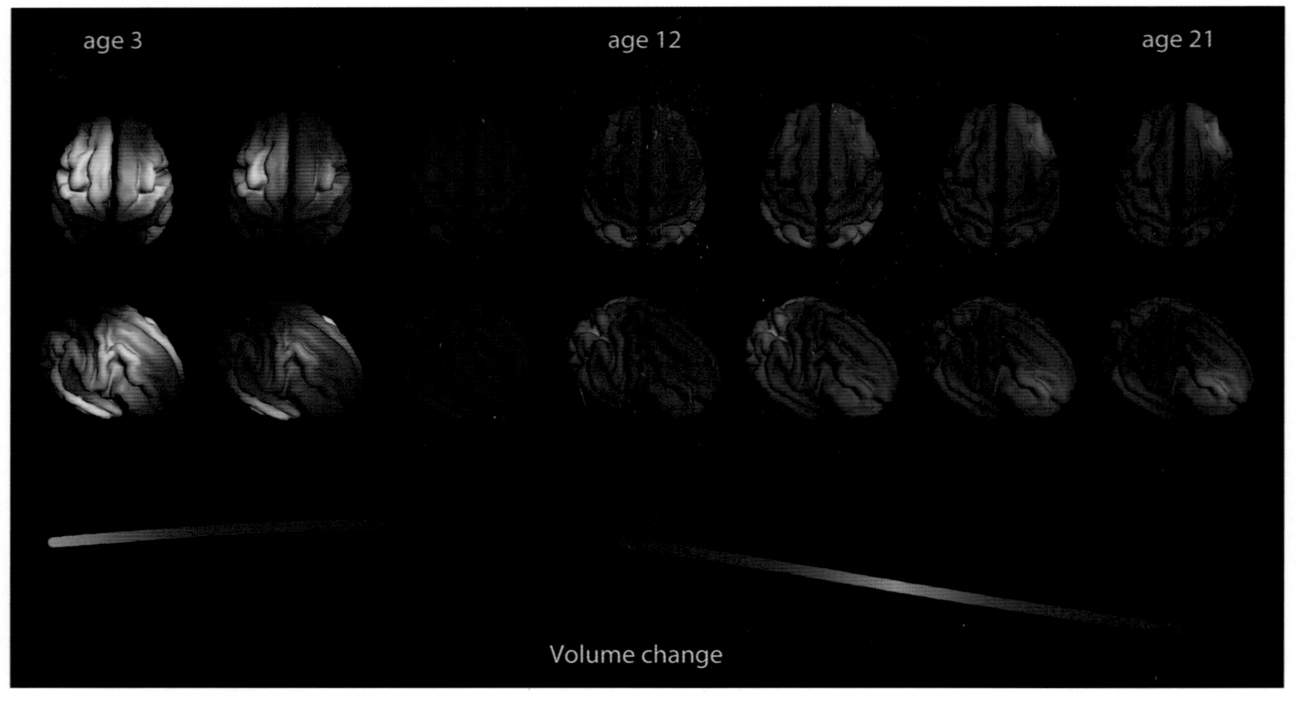

6. Top and right lateral views of change in cortical volume. Rate of change is shown from age 3 to 21, in 3-year intervals. Colour scales range from 2 per cent increases (yellow) to 2 per cent decreases (cyan) in volume per year. The bottom half of the image shows the developmental change of total cortical volume. This picture was generated with the open-access Pediatric Imaging, Neurocognition, and Genetics (PING) Data Repository, *Neuroimage* **124**, 1149–54 (2016). The displayed data include 1,234 subjects collected at ten sites.

Out of Control?

What happens if people fail to control their thoughts and/or actions? We have seen that frontostriatal circuits are important for certain aspects of cognitive control. It is not surprising that dysfunctional frontostriatal circuits can lead to serious mental disorders, often characterized by impaired cognitive control. Such dysfunctions are observed in conditions such as obsessive–compulsive disorder (OCD),[19] Tourette's syndrome,[20] and attention deficit hyperactivity disorder (ADHD).[21] Patients who suffer from these conditions have problems of control over their actions, thoughts, or both.

Early evidence for the relationship between intact neural networks and self-control comes from lesion studies. In 1996, Jordan Grafman from the National Institutes of Health led a study that tested veterans who had sustained head injuries during their service in Vietnam and healthy controls for measures of violence and aggression. The participants and their families answered questions about the subjects' aggressive and violent behaviours and attitudes. The results were striking: patients who had damage in the ventromedial prefrontal cortex (a region behind the eyes) had significantly higher scores on aggression/violence scales than healthy controls or veterans with brain damage in other regions.[22] The ventromedial prefrontal cortex seems to be of special importance for processes of self-control.

In Chapter 5, we briefly met a 40-year-old schoolteacher who suddenly became a paedophile, until a tumour in his brain was removed. He developed a sudden interest in child pornography and made advances towards his stepdaughter. The young girl told her mother, who informed the authorities. The man was removed from his home and had to undergo a rehabilitation programme for sexual addiction to avoid imprisonment. Later, he was expelled from the rehabilitation centre, because he molested nurses and other clients. Despite his strong desire to avoid prison, he could not control his urges. The night before his prison sentencing, the man went to the emergency department with complaints of a headache. He also developed balance problems and was sent for a brain scan. The scan showed a tumour in the right orbitofrontal cortex. After removal, he was able to complete a Sexaholics Anonymous programme and was allowed to return home. A few months later, he secretly collected pornography again and the headaches returned. An MRI showed that the tumour had regrown and after its removal the man returned to his original state.[23]

We met this man in the context of immoral behaviour, but it seems that his moral knowledge was intact. He tried to hide his activities because 'he felt that they were unacceptable'.[23] Even faced with the option of prison, he could not control his urges. This suggests that the orbitofrontal cortex—the region affected by the tumour—plays a key role in impulse control. Nonetheless, it has also been implicated in moral processing,[24] suggesting that self-control and morality are overlapping, if not tightly linked, mental processes.

This is only one anecdotal account of the serious consequences of a lack of self-control, but there are many more. In 1990, two American scientists, Michael Gottfredson and Travis Hirschi, developed a controversial theory. According to their 'General Theory of Crime', people with low self-control are more likely to commit a crime when given the opportunity. In their view, self-control develops during childhood and is influenced by parenting style. It then remains relatively stable in later life. Gottfredson and Hirschi argue that people with low self-control prefer to satisfy their immediate needs while paying little attention to long-term consequences, which makes crimes with immediate gains attractive to them.[25] A large number of research teams tried to test the theory and many have found evidence that self-control plays a role in criminal behaviour. Ten years after publication of the theory, Travis Pratt and Francis Cullen from the University of Cincinnati analysed the results of twenty-one studies, including over 49,000 participants. The studies included different ways of measuring self-control. Participants were from the general community, offenders, males, females, adults, juveniles, and from different ethnic groups. Overall, the authors concluded that low self-control is indeed 'an important predictor of crime', with a similar effect in different samples.[26] Despite this strong evidence in support of the theory, many have criticized the 'General Theory of Crime'. One of the major critiques is that the theory is too simplified and vague. For example, self-control is not a uniform construct, but Gottfredson and Hirschi do not clearly define it. The same is true for crimes, which can vary from minor offences to major crimes.[27] Nonetheless, many studies do support the importance of low self-control for criminal behaviour.

A study from 2003 highlights an even more dramatic finding: Eyal Aharoni and his colleagues from the Lovelace Biomedical and Environmental Research Institute titled their publication 'Neuroprediction of future arrest'. The team scanned ninety-six adult male offenders before they were released from prison. The participants performed a simple go/no-go task: when they saw the letter X on the screen, they had to depress a button as quickly and accurately as possible; when they saw the letter K, they had to withhold a response. Because participants see the letter X (go stimulus) a lot more often than the letter K (no-go stimulus), it becomes pre-potent to respond. Every response on a no-go trial counts as an error. The researchers tried to predict who of the offenders would be arrested again over the following four years, excluding arrest for minor parole or probation violations. They specifically looked at the activity in the anterior cingulate cortex, an area shown to be involved in error-monitoring.[17] Subjects with lower activity in the anterior cingulate cortex also made more errors on the go/no-go task. Strikingly, these subjects were also more likely to be arrested again.[28]

The results sound very impressive, and if they can be replicated we may have a useful tool for assessing programmes aimed at stopping reoffending. If criminals with low activity in the anterior cingulate cortex really are more likely to be arrested again, do they need special supervision or more extensive rehabilitation programmes? Before we jump the gun here, let us consider some major limitations of the study. The prediction of rearrest is not true for every participant: the risk is simply higher. The researchers divided the offenders into a group with high anterior cingulate cortex activity and one with low activity. The probability of rearrest was 31 per cent in the high-activity group and 52 per cent in the low-activity group. So indeed, the anterior cingulate cortex activity seems to be correlated with the risk of being arrested again. But how strong and reliable is this correlation?

The low-activity group, which included people with presumably poor self-control, were rearrested with a risk of 52 per cent. That means that about half of these people were not arrested again. If such a prediction of future arrest only applies to half the people in the group, are targeted programmes justified? We would need to invest large sums of money and the time of trained personnel and previous offenders. To use limited government funds wisely, we need a detailed cost–benefit analysis. It is certainly desirable to reduce reoffending because of its high cost to the individual and society. However, deciding who we want to put into these programmes based on such weak predictions might not be the most effective way. Future studies in bigger samples will have to show whether these associations can be stronger and more reliable.

But there is an even more important limitation: to make reliable predictions about future actions, the findings would need to be replicated in an independent sample. In this study, the authors already had information about rearrests, because the offenders had been followed up over a period of four years. The researchers used the same subjects for neuroimaging and frequency of reoffence. It would be useful to follow up this study with a new, independent cohort. If the brain activity of one group of offenders predicted the behaviour of an independent group, we would have stronger evidence of a reliable association. We need this replication to determine whether it is possible to make accurate predictions about rearrest. It would be helpful to develop highly accurate neuroimaging predictors of those who would be most likely to benefit from being offered a targeted programme.

While many studies do support the importance of self-control for criminal behaviour, they do not prove that it is the only important factor. People might commit crimes because they are desperate and see no other option. Maybe they want to impress their peers. Or maybe they are just bored. The truth is, there is no 'typical' criminal whose behaviour could be easily explained with a simple concept. Not all criminals have impaired self-control, and the opposite is true as well: not everybody with problems restraining themselves will commit a crime.

However, we do know that the ability to regulate ourselves is important and that failing to do so *can* have serious consequences. How would we treat subjects with a lack of inhibitory control, shown in brain scans? Let us take the example of sexual arousal. If a participant was unable to control sexual arousal and did not show activation in areas related to self-control, what would we do? Are they likely to commit a sexual crime and should they be obliged to register with the authorities? This kind of evidence could also be used after a crime was committed: if brain scans of sexual offenders show that they cannot inhibit arousal, are they really responsible for their actions? Or should this impairment be treated as a brain dysfunction and mitigating factor in court?

Can We Control Self-Control?

We know from experience in daily life and from experimental studies that there are people with problems of self-control. Earlier theories such as the 'General Theory of Crime' have assumed that self-control is a generally stable trait that varies between people.[25] But can it vary within the same person? Are there ways self-control can be impaired or improved?

One line of research indicates that self-control is indeed sensitive to manipulation. In 1996, Roy Baumeister from Case Western Reserve University and Todd Heatherton from Dartmouth College developed the theory of self-control as a limited resource. According to the model, the ability to regulate one's behaviour is much like a muscle, which can become exhausted after using it for long periods of time.[29] The researchers went on to design a series of experiments to test this theory. One very simple study was published two years later by Roy Baumeister and colleagues. Sixty-seven psychology undergraduates signed up for what they thought was an experiment on taste perception. Some of the participants entered a room filled with the smell of chocolate-chip cookies. There were two types of food displayed on a table in front of them: chocolate-chip cookies on one side and radishes on the other side. In one condition, the subjects were asked to take about five minutes to taste the chocolate-chip cookies, but not to eat any radishes. In another condition, subjects had to taste the radishes and avoid the cookies. Afterwards, the participants were asked to solve two puzzles, seemingly unrelated to the taste experiment. A third group of participants completed the puzzles right away and skipped the food part. All participants were told that they could take as much time and as many attempts as they wanted. What the subjects did not know was that the puzzle was designed to be impossible to solve. The researchers wanted to see how long the subjects were willing to try. The striking result: participants in the radish condition, who had to control themselves to not eat the tempting chocolate cookies, gave up significantly sooner than participants who ate cookies or who ate no food during the experiment. The authors concluded that these two seemingly different measures of self-control—resisting a tasty treat and not giving up on solving a tricky puzzle—are controlled by the same, limited resource.[30]

Many studies have tried to replicate this so-called ego-depletion effect. Martin Hagger from the University of Nottingham and his colleagues conducted a meta-analysis of eighty-three studies. The results suggested that self-control is indeed a limited resource. However, the authors add that these psychological studies cannot fully explain the mechanism by which self-control is depleted. They remark that mood, fatigue, and even blood glucose levels have an important influence on one's capacity to self-regulate.[31]

Hagger and his colleagues decided to take this one step further: they started a large replication attempt. They argued that the literature on self-control as a limited resource might suffer from publication bias—the general trend to publish positive results and to leave studies with negative results unpublished. To avoid this bias, the researchers created a Registered Replication Report. Before starting the study, they announced a plan for the experiments and agreed to publish their results no matter what the outcome would be. The same computer-based task was tested by twenty-three different research groups from many different countries in over 2,000 subjects. This task was developed by Chandra Sripada and his colleagues from the University of Michigan.[32]

It works like this: in the first part, one word at a time is shown on a computer screen. Subjects in the control group do quite an easy task—they press a button whenever the word contains the letter e. In contrast, subjects in the ego-depletion group get a harder task. They press a button only if the letter e is not next to or one letter away from a vowel. So they would press for the word 'trouble', but not for the word 'business'. In the second part, three numbers between 0 and 3 are presented. Two of these numbers are the same, one is unique. The participants identify the unique number. To make a response, they press one of three buttons. Their index, middle, and ring fingers are assigned to the numbers 1, 2, and 3, respectively. The subjects must now respond to the identity of the number, rather than its location. For example, they would see the sequence '2 1 2' and respond with their index finger (corresponding to number 1) rather than their middle finger (which would correspond to the location of the target). For the ego-depletion group, both tasks require a fair amount of effort. But does the first task drain self-control so much that it affects the second task? The large replication study with over 2,000 subjects showed little evidence of pronounced ego depletion. Only two of the twenty-three studies found that participants in the ego-depletion group performed much worse on the second task than the control group. Findings from the other laboratories did not support the idea of a strong ego-depletion effect.[33]

However, this study is not without critics, one of them being Roy Baumeister, who originally proposed the idea of ego depletion. The study team consulted him when they designed the study, but many of Baumeister's suggestions were not feasible for such a large attempt. The task had to be standardized and computer-based, to make sure it could be run in virtually the same way in all the participating laboratories. Having the smell of fresh cookies in the testing room was therefore not an option.

Baumeister criticized the study on the grounds that doing a task on a computer screen instead of with paper and pencil could have important influences on the task's difficulty. He stated that he would also have used a different procedure for the letter-e task: he would have trained all subjects on finding words with the letter e first to establish a habit before introducing a more complex rule. According to him, this would have made the task more challenging, because subjects would see an e and experience the impulse to respond.[34] The jury is still out on the concept of ego depletion, but it seems that the concept may need qualification, as the effect is not robust under different experimental conditions.

Michael Inzlicht from the University of Toronto and Brandon Schmeichel from Texas A&M University are also critical of the concept (in fact, both also contributed to the recent replication attempt). They agree that performing a self-control task at one point might influence the performance on a second self-control task. But according to the authors this is not necessarily due to depleted self-control. They suggest an alternative mechanism: a shift in motivation.

If you resist that tasty chocolate cookie after lunch, you might be more likely to eat a piece of cake after dinner. But does this necessarily mean that you have used all your capacity to self-control for the day: that you are now 'out of control'? Maybe your motivation changes? While you want to control your sugar intake after lunch, you might actively choose to have dessert after dinner—a conscious decision not to control yourself because you feel like you have earned a treat. If you do not even attempt self-control, can we conclude anything about your capacity to self-control? Inzlicht and Schmeichel argue that motivation is often overlooked in self-control experiments. The first task of self-control is usually hard work and participants might just not feel like working hard again on a second task. Maybe they would be able to, but they choose not to because they are not sufficiently rewarded for working hard.[35]

When Mark Muraven and Elisaveta Slessareva from the University at Albany paid their participants based on performance in their second self-control task, subjects performed equally well, no matter whether they had performed a first 'depleting' self-control task or not. It seems to matter how motivated the participants are to do well on the task. This motivation does not have to be money. In a different experiment, Muraven and Slessareva told the subjects that their performance on a problem-solving task would provide important insights into memory and could help to develop new treatments for Alzheimer's disease. Participants who performed an earlier self-control task did just as well on the puzzle as a control group.[36] No matter whether subjects are motivated by money or the belief that their actions are important for other people, increased motivation leads to improved performance on self-control tasks.

Moreover, motivation itself can be influenced by other factors. For example, the willingness to control one's actions is context dependent. You might relax and get a bit tipsy if you are enjoying a conversation with friends in a pub on a Saturday night. In contrast, if you are at your office Christmas party and sit next to your boss, you might be more likely to control your alcohol intake. The social context determines your motivation for self-control, not necessarily your ability to practise it.

The theory of self-control has been further developed by Todd Heatherton and his colleague Dylan Wagner from Dartmouth College. In 2011, the pair suggested a balance model of self-control. They argued that successful self-control relies on the balance between two systems: the prefrontal cortex for cognitive control and deeper brain regions involved in reward and emotion. A failure occurs when either the prefrontal cortex is impaired or when the reward/emotion system becomes too strong.[37] In other words, when you eat a piece of chocolate cake even though you are on a diet, this might be because your prefrontal cortex cannot control your reward centres well enough or because the delicious smell of the cake activates your reward centres too strongly. Or maybe it is a combination of both.

Nora Volkow, the current director of the National Institute on Drug Abuse in Maryland, showed this effect very impressively in cocaine addicts. Video clips that showed people buying, preparing, and smoking cocaine were shown to twenty-four cocaine abusers. In some of the trials, the subjects had to suppress their craving for cocaine; in others they were told not to do so. When the subjects actively inhibited their craving, they showed lower activation in areas related to reward processing, namely the orbitofrontal cortex and a part of the striatum. This decreased activity was also associated with higher activation in frontal areas, involved in cognitive control.[38] This evidence suggests that it is indeed a balance between control and reward areas that makes self-control possible.

But what happens during depletion of self-control? A study lead by Dylan Wagner addressed this question in chronic dieters. A total of thirty-one women who fulfilled criteria for chronic dieting completed fMRI tasks. First, the women watched a short documentary on sheep. During the video, words appeared at the bottom of the screen and moved to the centre. Half of the women were allowed to read the words; the other half had to inhibit reading the words and focus on the documentary. This might sound easy, but these distracting words are actually fairly hard to ignore and this paradigm is well established for depletion of self-control. After watching the sheep documentary, the women completed a second task. They were shown pictures of tasty food such as desserts, or neutral pictures showing people or landscapes. For each picture, the dieters simply had to decide whether the picture showed something indoors or outdoors, to make sure that they stayed alert.

The interesting analysis of this study was the difference between the two groups of dieters. The women who had previously performed the self-control depletion task showed higher activity in the orbitofrontal cortex in response to food cues. Among other processes, this brain region behind your eyes is involved in reward processing. In addition, these women also showed reduced connectivity between the orbitofrontal cortex and the inferior frontal gyrus, a part of the prefrontal cortex involved in cognitive control.[39] This evidence suggests that when self-control is depleted, a chocolate cake seems more attractive and it is harder to control your craving.

The chocolate cake might seem particularly irresistible when you are stressed. Silvia Maier and her colleagues from the University of Zurich showed the remarkable effects of stress on decision-making and brain activation. The participants were fifty-one healthy men who made an effort to eat healthily and exercise, but who also enjoyed junk food and consumed on average seven to eight items of junk food per week. In the beginning of the experiment, the men rated 180 food items in terms of healthiness and tastiness. Before going into the scanner, about half the men completed a stress test. They had to put their hand into ice-cold water for three minutes while looking into a camera that recorded them. The other participants put their hand into warm water and they were not videotaped. Afterwards, the men made choices about food items while undergoing fMRI. They chose between two food items and were told that they were expected to eat one of these selected items, randomly chosen after scanning.

When making their choices, the stressed participants were more likely to make their choice based on the tastiness of the item compared to the control participants. Especially when the difference in taste between the two items was big, the stressed subjects tended to choose a tasty, unhealthy item over a healthier, less tasty item.[40] Basically, the chocolate cookie would beat the radish for stressed men. This effect had already been established by then and was not surprising.[41] What made the study special was that the researchers looked at the difference in brain activation between the stressed and non-stressed participants. In the stressed men, the researchers found that the relative tastiness of the food was more strongly correlated to activity in the ventral striatum and amygdala. Both are deep brain structures that have been shown to be involved in reward processing and assigning a motivational value to a stimulus.[42, 43] Moreover, the stressed men also showed reduced connectivity between two prefrontal regions which are involved in self-control.[40] This study demonstrates the dramatic effect of stress on brain function and on our ability to regulate ourselves. Next time you crave that chocolate cake while being on a diet, maybe indulging instead in a relaxing activity, such as yoga or going for a walk, is a better option.

Are there any other ways in which self-control can be improved? Another way of avoiding the tasty chocolate cake might be to actively think about your future. Tinuke Oluyomi Daniel and colleagues from the University at Buffalo tested this idea with twenty-six overweight or obese women. Their hypothesis: in order to reject an immediate reward in favour of a future reward, participants have to be able to think about future consequences. In other words, if you want to say no to the chocolate cake, it would help to imagine 'the healthier future you'. About half the women completed a task in which they had to think about their future by listing possible future events. The other women read a travel blog describing vivid events. Next, the women had to rate the appeal of high-calorie foods, such as meatballs, sausages, and cookies, but were only allowed to look at them. This paradigm was used to increase their craving. Afterwards, the women had unlimited access to the foods for fifteen minutes and were asked to rate the taste and texture. What the women did not know was that the researchers did not care about these ratings. They were only interested to see how much food the women would consume in the fifteen minutes.

Thinking about the future had a dramatic effect on the overweight and obese women: they consumed on average 300 calories less than the group who thought about the travel blog.[44] That's about 1.5 chocolate cookies or five medium-sized meatballs less.

So actively thinking about your future can influence what you do in the present. A busy student might find it easier to stay home and study rather than going to a party if he or she actively thinks about future rewards: a great degree, a desired job? It is all about perspective.

In addition to actively thinking about the future, there might be other ways to improve self-control. Mark Muraven investigated whether training could have beneficial effects. He let ninety-two adults practise a task for two weeks. The participants had to either avoid sweets, squeeze a handgrip twice a day, which caused physical discomfort, solve maths problems for a few minutes each day, or keep a diary about any acts of self-control. Avoiding sweets and squeezing the handgrip for as long as possible were tasks designed to train self-control, because they require the subjects to inhibit their urges. In contrast, spending a few minutes a day solving a relatively simple maths problem or keeping a diary should not require large amounts of self-control. Muraven found that the people who had cut back on sweets or squeezed the handgrip showed improved self-control after two weeks compared to the other subjects. Importantly, all subjects were told that they were training their self-control. We can therefore rule out that the anticipation of improved self-control had a significant effect. Only people who really did engage in self-control training benefitted from it.[45]

Studies like these show us that self-control does change depending on our experience or environment. It is not a personality trait set in stone but rather a resource which can be depleted but also increased. This has important implications for the treatment of disorders and unhealthy behaviours characterized by low self-control. Mark Muraven wanted to find out if he could help smokers to quit by training their self-control. There were 122 smokers who participated in his experiment, performing one of the tasks we have just described—involving sweets, handgrips, maths, or diaries. All the participants wanted to quit smoking. After two weeks of daily training, they were told to quit smoking. For the next twenty-eight days, the participants reported back to the study team every day saying whether they had been smoking again. During this time, their breath was also analysed on four occasions to verify whether they had been smoking. The impressive results: subjects who had practised the self-control tasks for two weeks were about 1.5 times more likely to successfully stop smoking.[46]

Such results show us that self-control training could have great benefits for patients who suffer from conditions of low self-control. If a simple training of two weeks can have such effects, would long-term training prove successful for people with substance-abuse problems, pathological gambling, or obesity? Science is only beginning to indicate that self-control training has beneficial effects on brain circuits and behaviour[49] and the results look promising. The idea that cognitive training has strong effects on behaviour is intriguing, especially since these interventions carry little risk but great potential benefits.

How Much Control Do We Really Have?

Given that we are in a chapter about self-control, let us take this one step further: do we have any control at all?

Neuroscience has recently started to touch on a fundamental philosophical question: do we have free will? A groundbreaking study was published in 2008 by the group of John-Dylan Haynes from the Max Planck Institute for Human Cognitive and Brain Sciences. They invited fourteen participants to watch a stream of letters while undergoing fMRI. When they felt the urge to do so, the subjects were required to press one of two buttons. It was entirely up to them when to press and which button to choose. The participants also indicated which letter was shown when they made the decision to press. Therefore, the researchers could identify when the subjects made their conscious decision to press. Most of the subjects made the conscious decision within the second before pressing the button. But the researchers tried to answer a different question: could they find a brain region which predicts which button will be pressed before the subject made the conscious decision? In other words, is there a brain region that 'knows' or 'decides' what will happen before the subject does? The researchers argue that they have found such a brain region. It is part of the anterior prefrontal cortex, a region just behind the forehead. The scientists found that the signal predicting which button will be pressed can already be detected up to seven seconds before the subject made the conscious decision to press.[47] This was an unexpectedly big time difference. The data suggest that decision-making is very complex and involves unconscious processes before entering awareness. The interpretation of these results is difficult. One important limitation of the study is that the model did not work perfectly. Using the signal from the anterior prefrontal cortex, the researchers could only predict the response accurately three out of five times. While this is certainly higher than chance, it is far from being a perfect predictor.

A later study conducted by Itzhak Fried from the University of California and his colleagues used a different method. Patients with epilepsy were undergoing a surgical treatment. They had electrodes implanted in their brains to find the focus of their seizures. However, the researchers also used the electrodes for an experiment. The patients watched an analogue clock with a rotating hand. Whenever they felt the urge to stop the clock, they were required to press a button. Afterwards, they pushed the hand back to the spot where it had been when they made the decision to press. Using recordings from the electrodes, the team was able to predict a response correctly four out of five times up to 700 ms before the conscious decision. These recordings were made from the supplementary motor area,[48] a region on the upper part of your brain involved in motor control. This remarkable result was probably possible because the recordings from single electrodes are far more sensitive than fMRI signals. The study provides more evidence of an unconscious decision-making process.

If a decision is being made in a part of your brain before you become aware of it, is this still your decision? You could argue that it is still a part of *your* brain that makes the decision, so ultimately, you. The study only shows that some processes take place before you become aware of a decision: it cannot definitely prove the absence of free will. At this point, the debate turns into a philosophical rather than a scientific one. We will leave it up to you to make up your own mind.

Gaining Control

We have seen that self-control is an important resource. Your ability to wait for a second marshmallow at age 4 does not necessarily predict the course of your life, but it is still a remarkable and essential process. Losing control over your thoughts, urges, or feelings can have serious consequences such as criminal behaviour, unhealthy habits, or even serious mental disorders. But does that make you less responsible for these acts?

It is helpful to be aware that stress or repeatedly failing to exert self-control when appropriate could impair our ability to regulate ourselves. Having tight deadlines or trying to ignore your Facebook notification at work could make you more vulnerable to eating the tasty cake at lunchtime, no matter how sincerely you have sworn to lose weight. But repeatedly practising small acts of self-control could make you stronger and help you to resist temptation. Is it time to take control?

7

SHOW ME YOUR BRAIN AND I KNOW WHAT YOU BUY?

It is the most expensive advertising space on TV: America's Super Bowl. Companies are paying some £3m for a 30-second commercial, and the prices are increasing every year. This does not include the actual design and production of the spot, only the right to show it to one of the biggest TV audiences.

Even though this is an extreme example, advertising costs companies large amounts of money and they want to invest these sums wisely. It is not surprising that they would look for tools to test their ads, to see how consumers react before companies make financial commitments. These tools used to be basic questionnaires or focus groups, but these are not perfect. What a person explicitly says in a questionnaire or focus group might not be an objective account of his or her preferences and feelings. People could be biased or they could give vague answers.

They might decide to keep their real opinions to themselves or they might not even be aware of their true preferences. Some processes of decision-making could be subconscious, so do people really understand what they want? Will neuroimaging reveal information about our purchase desires that we would rather not admit to, or were even unaware of?

There is a demand for more objective and predictive measures to help companies understand how consumers make their decisions and which products will be successful. A new hope is the application of neuroscience to marketing—so-called neuromarketing or consumer neuroscience. The worlds of science and business could meet at any of three different stages: when the idea of a product is being developed, when the finished product is evaluated, or when a marketing strategy is tested.

When is the application of neuroscience to product development and advertising justified? It really only makes sense if these techniques can tell us something we would not otherwise know. If we do not want to rely on people's explicit answers, we could also measure their physiological response by monitoring skin conductance or heart rate. Is neuroscience really superior to other body measures? Can it add something to the information we get from simple questionnaires?

Beyond Traditional Measures

This question was addressed by Vinod Venkatraman from Temple University and his colleagues. Their study was funded by the Advertising Research Foundation, a non-profit association based in New York City. Among its members are major companies, advertising firms, media companies, and academic institutions. The hope of the study was to find the most promising technique for market research—one that can reliably predict the future success of a product. They tested six different techniques and compared their predictive value: self-reports, implicit measures (similar to the Implicit Association Test you read about in Chapter 3), eye tracking, biometrics (heart rate and skin conductance), electroencephalography (EEG, which uses electrodes to detect changes in electrical fields in your brain), and fMRI. The participants watched thirty-seven TV ads while their responses were measured with a selection of the listed techniques. The researchers then used these data to predict the success of the advertisement and compared it with the actual advertising outcome.[1]

By far the best predictors were the traditional self-report measures. Asking subjects directly how much they like the ad and if they intend to purchase the product still seems to be the best predictor of market success. It is worth noting, however, that 186 participants completed these traditional measures. In contrast, only twenty-nine participants completed eye tracking, biometrics, and EEG, and thirty-three participants were part of the fMRI group. These are fairly standard sample sizes for the respective techniques, but we have to consider these differences in group size when we interpret the results. For example, if more people had participated in the EEG study, these data may have been more informative. An advantage to neuroimaging may be the gain of marketing information in relatively small groups, which would in part offset the expense of neuroimaging costs.

So is there any hope for the emerging field of neuromarketing? The researchers wanted to see if there are any techniques beyond the traditional measures which could predict market success. They did find one: the only technique which was able to improve the prediction was fMRI. Moreover, there was only one brain region which proved to be significant: the ventral striatum.[2] This is a deep area towards the centre of your brain which is involved in the prediction of reward.[3] The fMRI data improved the prediction by only about 5 per cent, but bigger sample sizes and more sophisticated methods of data analysis might make these data more useful. Also, even a small improvement might be helpful when a company designs a multimillion-pound ad campaign.

This was the first study to directly compare the value of traditional and newer measures in predicting ad success. Despite its limitations, the study suggests that fMRI has something to offer for marketing and that it is worth investigating this further. And other support for this application of fMRI was to follow.

The Use of fMRI in Marketing Research

It all started with Pepsi and Coca-Cola (Coke). A study led by Samuel McClure and Jian Li from the Baylor College of Medicine looked at the difference between the two soft drinks. Pepsi and Coke are very similar in composition and taste. In blindfold taste tests many people cannot identify which one of the two they are drinking. Pepsi even started an ad campaign in the 1970s called the 'Pepsi challenge', in which participants at shopping and other public spaces blind taste tested the two beverages. This ad was meant to encourage people to focus on the taste of the beverages rather than the cultural associations.

Despite these similarities in taste and composition, Coca-Cola holds the bigger market share. Why is one brand so much more successful than the other? The researchers invited participants to taste Coke and Pepsi while undergoing fMRI to find out. Beforehand, the participants performed a blind taste test and chose which drink they preferred. Pepsi and Coke were chosen equally often. Afterwards, the drinks were administered in the fMRI scanner and the participants' brain activity was measured. The study had an interesting twist: sixteen of the participants were not told which brand they tasted in the scanner and they tasted both Pepsi and Coke in a random order; another sixteen subjects only tasted Coke, but in some cases they were told that it was Coke, in others they just saw a light. The researchers used this design to see how the pure experience of taste differed from the brand-cued experience. The results were remarkable.

When subjects were not told what they were drinking, the researchers could see a relationship between their blindfold preferences from the taste test and brain activity. Specifically, the activity in the ventromedial prefrontal cortex was correlated with blindfold preference.[4] This brain region above the eyes is important for evaluating reward, especially the subjective value of reward.[5] When the ventromedial prefrontal cortex had a stronger activation in response to Coke than Pepsi, the subjects were also more likely to prefer Coke over Pepsi in the taste test. Essentially, your brain shows which drink you really prefer.

What happened when the participants were brand-cued? Remember, they tasted only Coke in this experiment, but sometimes they were told it was Coke, sometimes it was left open. In this case, the sensory information was exactly the same—no matter whether the drink was labelled or unlabelled, they tasted the same beverage. But the fMRI data revealed that there was a strong difference in brain response between the two. Knowledge of the brand produced increased activation of the hippocampus and dorsolateral prefrontal cortex. The hippocampus is a deep brain region involved in the formation and recall of memories, among them autobiographical memories.[6] The dorsolateral prefrontal cortex is located above the hairline towards the side of the brain. As we have seen, this region is important for cognitive control.[7] Knowing which brand of drink you are tasting could bias your perception, because you associate the brand with a cultural context.

This was one of the first studies to show how our perceptions are biased. But apart from branding, how else are our perceptions influenced? Imagine that you are invited to a dinner party and you go to the supermarket to buy a bottle of wine. You are in a rush, so you do not have time to read reviews or ask your friends for advice. How do you chose a good wine? One easy way might be to pick a wine by price—surely the cheapest wine will be horrible while the most expensive one is bound to be excellent?

A study from 2008 suggests that the price really does matter: it can dramatically change how we judge a wine. Hilke Plassmann from the California Institute of Technology and her colleagues let twenty participants taste red wines in an fMRI scanner and decide how much they liked them. The subjects were told that the bottles cost $5, $10, $35, $45, and $90 respectively. But here is the catch: the participants did not know that they were only tasting three different wines, the $5 and $45 wines and the $10 and $90 wines were the same. The intriguing finding was that when the price of the identical wine was increased, subjects reported that they liked the wine significantly more. The \$90 wine was rated about twice as pleasant as the \$10 wine, despite being exactly the same wine. The prices were also associated with different brain activity. Both the \$90 and \$45 wines led to higher activation in the medial orbito-frontal cortex than their cheaper counterparts.[8] This region behind the eyes is associated with the experience of pleasure. For example, it is activated when viewing a beautiful painting or hearing a pleasant piece of music.[9] A 'good' wine seems to have a similar effect, as the study suggests. Importantly, primary taste areas which are involved in the basic sensory perception of the wine did not show a difference between the prices.[8] The price probably modulates your perception on a higher cognitive level. If you choose the most expensive wine for your dinner party, it might not be objectively tastier, but you could still experience it as more pleasant than cheaper wines. Decide for yourself if this is a wise investment.

Studies like these could prove very valuable for companies when they decide on pricing strategies. Goods associated with luxury might be more successful when sold at a higher price, because both the expectation and experience of pleasure are increased. But can neuromarketing inform more than branding and pricing decisions?

According to a study by Gregory Berns and Sara Moore from Emory University, neuromarketing could directly predict the success of a product. The researchers wanted to see if they could predict the sales of songs by playing them to teenagers while recording their brain activity. Short clips were played to twenty-seven teenagers aged 12 to 18. The songs were from all genres—from country to metal—by unsigned or relatively unknown artists. The teenagers only heard songs from their favourite genres and rated how much they liked the songs. Surprisingly, these subjective ratings did not correlate with the songs' market success over the following three years. Asking teenagers these simple questions about songs might not be very useful for the music industry. But was there hidden information in the brain?

The commercial success of the songs was correlated with the teenagers' activity in the nucleus accumbens.[10] This deep brain region is involved in the processing of rewarding, pleasant stimuli.[11] The authors used this brain activity to classify a song as a future hit or non-hit. When a hit was defined as having 15,000 or more album sales, the model could correctly identify 80 per cent of the non-hits, but only 30 per cent of the hits.[10] However, it is a matter of definition what makes a hit a hit. A 'gold' album has to reach at least 500,000 sales, but only a few of the songs in the study reached the 'gold' criterion (probably because they were taken from relatively unknown artists). We would need songs with a wider variety of market success to see how useful brain imaging data really are in predicting sales. But how would we know which songs to include? Is this not the very aim of the study itself? You can see that neuromarketing has to overcome technical difficulties before it can be used in a sensible way. Considering the study on 'brand' we have described, it may be that well-known artists or 'brand' artists would have produced effects in different brain areas, such as the hippocampus, although there is evidence to suggest that real people and brands are not processed in the same way.[12]

This study may make it sound as if traditional measures are a lost hope for the music industry. After all, the subjective ratings did not prove to be useful for predicting sales numbers. But we have to consider that the sample size was small and the questions were very simple. The subjects were just asked about how much they liked the song and how familiar it was. A more detailed interview in a larger cohort might have been more useful. Also, teenagers only represent a subset of the market. The songs are bought by people of all ages and with different social and economic backgrounds. Asking a small subset of teenagers about their impressions may not be enough to predict market success. However, the fact that the adolescents do not represent the whole market makes the neuroimaging findings even more intriguing. If such a selected sample provides information about the entire market, this could prove very useful. Maybe the songs were only slightly different—these differences could be too small to be picked up in simple questions, but these subtleties might be reflected in brain activity.

The Future of Neuromarketing

Exactly this 'hidden information'—the subtle differences in perception that participants are not even aware of—is what makes neuromarketing so attractive. So far, we have only limited data to judge whether we can really reveal this information and use it in a beneficial way. Nonetheless, there are already countless neuromarketing firms that cooperate with advertisers. They promise to improve sales of a product by peeking into the inner workings of the consumer brain.

Unfortunately, these neuromarketing companies are not bound by academic standards. They usually do not publish their findings in peer-reviewed journals that are accessible to the scientific community. And they might have good reason not to—if a certain way of analysing neuroimaging data proves especially useful for marketing, this would be a major sales argument and advantage over other firms. If a neuromarketing company wants to keep its competitive edge, it might be damaging to share its insights and methods with the world. There could already be a brilliant way of predicting market success of a product or of perfecting its design based on neuroimaging data. How can we know?

However, we think it is unlikely that such a breakthrough has occurred yet, as neuroimaging suffers from some important limitations. Being in a noisy fMRI tube is not the same as casually strolling through a shopping centre and deciding which train set to get for your niece. We cannot directly infer consumer behaviour in the real world if the study situation is not realistic. fMRI along with most other neuroimaging methods does not allow a real-world monitoring of purchase decisions. Even portable EEG caps require the subjects to be as still as possible, ideally in an isolated room.

One point that is especially important for businesses is the financial cost. fMRI, along with most other neuroimaging methods, is quite expensive. When the University of Cambridge bought a state-of-the-art fMRI machine in 2006 it cost around £1.4m. Currently, the cost of running a 1-hour experiment is £620, with a significant amount of this money being spent on maintenance and the trained personnel who operate the scanner. Importantly, these reduced prices are charged by a university which is concerned only with the running costs of the equipment. If a neuromarketing firm were to buy an fMRI scanner and charge their commercial clients to do experiments, these commercial prices would probably be higher. To be truly useful and to justify such an investment, the technology would have to be superior to traditional measures. For example, supermarkets can collect data on consumer behaviour via loyalty cards, and online purchases can be tracked relatively cheaply and easily. These large data sets enable companies to target their marketing to the individual. It is yet to be proved that neuromarketing can overtake these methods, although they might also be used in combination.

Another major limitation is the one of reverse inference.[13] Essentially: 'if A, then B' does not necessarily mean 'if B, then A'. If your boss smiles at you and then you get a raise, this does not mean you will get a raise every time your boss smiles at you in the future. Similarly, one study might show that the cognitive process A activates brain region B. If another study now finds that brain region B is activated, this does not prove that process A was involved. Region B might be involved in many different processes, including processes C, D, and E. You get the idea.

The fallacy of reverse inference led to a public dispute in 2011. A branding consultant published an article in the Opinion Pages of the *New York Times* entitled 'You love your iPhone. Literally'. In cooperation with a neuromarketing firm he carried out an fMRI study with sixteen participants. They listened to and watched a ringing and vibrating iPhone and their brain activity was recorded. Here is what the author of the article wrote about his study:

> most striking of all was the flurry of activation in the insular cortex of the brain, which is associated with feelings of love and

> compassion. The subjects' brains responded to the sound of their phones as they would respond to the presence or proximity of a girlfriend, boyfriend or family member.[14]

Sounds like a remarkable finding? There is only one problem with this interpretation: the insular cortex is involved in a variety of processes. In fact, a meta-analysis of over 3,000 neuroimaging studies showed that the insular cortex is one of the most commonly activated brain areas.[15] It seems to be involved in a variety of tasks that require attention towards a salient stimulus. So the interpretation that people love their iPhones is not justified, based on the findings. It even led to forty-five neuroscientists writing a letter to the *New York Times* explaining why they felt the claims were not scientifically valid.[16]

Instances like these should remind us that the brain is not a simple organ. Cognition works by activating a variety of brain regions which communicate with each other and modulate each other's activity. One brain region can be part of many circuits and involved in different processes. This is why we do not have a 'brain centre of love' or the 'morality region'. And only if a region is consistently shown to be activated by a certain process as shown by different researchers with different study designs, only then can we assume that it is involved in a certain process. Indeed, scientists usually confirm findings by using several different techniques which all converge on the same answer, rather than relying on only one single experimental design.

However, we forget one important point here. Instances of reverse inferences may be very unsettling for the academic community. But to build a valid model that predicts the market success of a product, we do not need any interpretations or inferences of the brain network. If a computer algorithm could show that a certain circuit in the brain can predict the sales of a pop song with 90 per cent accuracy, do we really need to understand what this circuit does in the brain?

Many of us would very much like to know what it does, and researchers could aim to find this circuit's function. But for a pure business application, such a black box would not be a problem. As long as it predicts the market success well enough, businesses do not necessarily need to understand the neuroscience behind it. This may sound very promising for some and deeply unsettling for others. But how many phenomena do we truly understand anyway? In the case of a number of drugs that are routinely given to patients, we do not fully understand which of their many actions are responsible for the beneficial treatment effect. They just work, and that is enough to justify their application. Why should neuromarketing be any different?

Is Neuromarketing Ethical?

Let us assume that neuromarketing will prove successful one day. Different versions of Super Bowl ads could be shown to people, and their brain activity would tell the director which one is most successful. Maybe we could 'design' the perfect political candidate, because neuromarketing tells us which haircut, outfit, and speech has the optimal outcome. Or could neuromarketing help to create the ultimate obesity factor—a food item that is so addictive that we cannot stop eating it?

Is there a fundamental difference between conventional marketing and neuromarketing? If it proves to be true that there is a wealth of hidden information in our brains that can be accessed and used for marketing purposes, new ways of manipulating the consumer might open up. But what about free choice? No matter how well made an ad or how delicious a certain food, you are not forced to purchase and consume it. Ultimately, it will still be your decision what you put on the dinner table and which political party you elect. But your choices could be influenced and manipulated by clever marketing.

This is especially worrying when the ability to self-control or to make good decisions is not at its optimum. That is true for all of us every now and then: we could be more susceptible to targeted marketing after a stressful day. And what about children, whose brains are still developing and whose cognitive functions have not reached their full potential? Do we want to allow neuromarketing to manipulate them beyond the current market practice?

Some might fear that neuromarketing could be used against us. Companies could make us pay even more for their products because neuroimaging helped them to identify the ideal pricing strategy. But it could also be used in a mutually beneficial way. If pop songs, films, or foods are designed to be even more pleasant, don't we all benefit from this? Maybe neuromarketing could be used to make healthy, nutritious foods more attractive and help to fight obesity.

Perhaps the biggest concern is that there are few standards and guidelines for using neuroscience for business. While academic institutions have clear guidelines as to the conduct of studies and how to avoid harm and adverse events for participants, these rules may be lacking in business. If a researcher wanted to conduct a study based at a university, an ethics committee would have to evaluate the study and see if there are potential harms for the subjects—essentially, if it is ethical to conduct such a study. The academic researcher could not conduct the study without approval from the ethics committee. In contrast, if a company buys an fMRI machine and decides to use it for neuromarketing purposes, there are few regulations and supervisions. In the US, the Food and Drug Administration (FDA) regulates the use of fMRI machines because they are defined as medical devices.[17] However, such regulations can only ensure that the machines are operated with care, not that the study protocols are ethical. The companies could show deeply traumatizing scenes to participants. They might not be transparent about the aims of their research and subjects could end up revealing information they would have preferred to keep private.

And, of course, there is the issue of discovering a brain abnormality without looking for it. Most universities have clear protocols in place about what to do when a participant's brain scan shows an abnormality. Sometimes this does not lead to any symptoms and does not require treatment; sometimes tumours or other serious conditions are discovered. Usually a neuro-radiologist would look over the scans and, if necessary, the subject would be informed without giving a medical diagnosis (because fMRI is not a diagnostic instrument and researchers are usually not qualified to make a diagnosis), although the protocols vary between institutions.[18] However, businesses might not have such protocols in place. How can we make sure that such incidental findings are dealt with in an ethical manner?[19, 20] Which regulatory body would have authority over neuromarketing firms?

To conduct research responsibly, neuromarketing firms should adhere to a code of ethics. The subjects should be informed about the aims and risks of the research and give consent before any data are collected. Debriefing procedures may also be necessary. This is especially important for vulnerable groups, such as children, patient groups, or other legally protected groups. The information should be treated as confidential and not be used to stigmatize the subjects. Moreover, neuromarketing companies should report their findings accurately and not overstate them. If they do not publish in scientific journals, their studies cannot be reviewed critically by other experts. We would not have a chance to assess their methods and decide if the studies themselves were ethical. Also, other scientists would not be able to examine the findings carefully and in detail. This would give companies the freedom to make invalid claims which could create false hope or anxiety in the general public. Moreover, scientific findings are often complicated and nuanced. If they are oversimplified and presented as absolute truths, this could give the public the wrong impression of real scientific research. This risk is especially high when these companies have a strong commercial interest in making neuromarketing successful.

All things considered, neuromarketing probably has great potential. If it really can inform marketing beyond traditional measures, companies would be able to invest their money more wisely. Their findings could contribute to our knowledge about the brain and how it works in the context of decision-making. There could also be benefits for the individual. Products could be designed to be more pleasant, and entertainment options could be even more enjoyable. To achieve these goals in a responsible manner, neuromarketing firms should follow strict ethical guidelines. We should have a public discourse about how far we want neuromarketing to go—the time is right for a discussion of its risks and benefits.

8

WHERE DOES THIS LEAVE US?

The development of neuroimaging has transformed the field of neuroscience. For the first time, we can see inside the living human brain safely and non-invasively. Most of us know somebody who has had an MRI scan for medical reasons, but the technique can be used for other very important and exciting applications. We may hope to understand some of our most basic emotions, motivations, and behaviours with the help of neuroimaging. Accumulating such knowledge will allow us to develop useful applications.

In this book, we have discussed some of the exciting findings from the use of fMRI and how they can be used to explore human behaviour. As well as the benefits, we have also highlighted some of the risks and limitations of the technique. Some critics go further: fMRI has been described as a modern version of phrenology, the popular nineteenth-century movement whose advocates argued that the shape and size of regions of the head corresponded to particular personality traits. Phrenology was later discredited as pseudo-science. Comparing fMRI to phrenology addresses a possible limitation of some neuroimaging studies: the (false) idea that cognitive processes can be located in distinct brain regions which are exclusively concerned with this process. We have seen in this book that complex processes cannot usually be traced back to simple brain regions which do not serve any other purposes. Morality relies on areas also implicated in emotional and social cognition. There is no uniform centre of self-control. And neuromarketing has yet to discover the area which will predict sales of all products. All these processes usually rely on a network of brain areas which are interconnected and active in a variety of tasks. This is sometimes forgotten when researchers interpret their results.

However, when scientists use language carefully and keep the vast complexity of the brain in mind, fMRI is far from phrenology. It can reveal new and intriguing information about the human brain. For example, a recent study by Jack Gallant's laboratory at the University of California mapped the semantic atlas of the cortex. Their seven subjects listened to more than two hours of stories while undergoing fMRI. The analysis showed which areas of the system respond to which concepts—for example, which area of the cortex responds to colours and which one to numbers. These semantic networks represent a sum of our verbal knowledge and how we have understood the relationship between words and concepts. A fascinating finding was that these semantic maps were very consistent across individuals, possibly because they were raised and educated in similar environments. If this lack of large individual differences proves to be true, we could learn something very important from individuals who do differ from the norm. Maybe we could detect language difficulties or superior semantic organization? Using fMRI for such visionary approaches shows us how far we have come from the days of phrenology (see Plate 5).

An important aspect of neuroscientific research, as in all scientific research, is replication. Sometimes scientists try to confirm the findings of an experiment with their own study, but do not get the same results. Part of the problem is that researchers approach topics such as morality, empathy, racism, and deception with different experiments. Changes in the experiment could have dramatic effects on the participants' perception, which might be why many studies cannot be replicated. It would therefore be useful to have a standardized set of tests than can be validated with large cohorts and used in healthy subjects and patient groups at a reliable standard. Such sets of tests already exist for cold, or non-emotional, cognition—one example being the Cambridge Neuropsychological Test Automated Battery (CANTAB). However, we need improved, validated tests of social, emotional, and moral cognition, termed 'hot' cognition.[1] A set of tests of 'hot' cognition, called EMOTICOM, is currently being developed by Rebecca Elliott, Barbara Sahakian, Trevor Robbins, Jonathan Roiser, and Mitul Mehta.

Despite the limitations, brain science has seen some great advancements. Neuroscientists can now carefully interrogate the complexity of the mind using neuroimaging. These exciting techniques can provide novel insights into how our brains work and help us understand why we act the way we do. For example, neuroimaging has provided evidence that our brain development does not stop in childhood. We now know that some brain areas develop until young adulthood, because subjects were scanned repeatedly during their development. Studies led by Jay Giedd and Nitin Gogtay from the National Institutes of Health have used structural MRI to look at these developmental changes.[2, 3] Plate 6[4] shows changes in the volume of grey matter, the tissue that makes up most of our cortex, from early childhood to young adulthood. You can see that many of these regions change dramatically during development.

In this book, we have discussed some possible applications of neuroscientific findings. Screening for terrorists at the airport with mind-reading techniques or using fMRI for lie detection in court are possible scenarios of the future. They are not reasonable and justified yet, but many of the current limitations could be overcome by advances in technology. More precise neuroimaging techniques could give us a better insight into brain functions. More sophisticated methods of analysis could make noisy data more meaningful. We have seen how promising machine learning algorithms are, in which the computer builds a model based on the data and learns from the input of future data. The researchers no longer need to know the perfect solution to a question and program it. In machine learning algorithms, the computer can figure out the best solution itself and get better with practice. The potential of machine learning has already been identified: recently, Google paid the rumoured price of about £400m to acquire the artificial intelligence company DeepMind Technologies, which has developed some advanced machine learning algorithms.[5] With such technological advances, new applications of neuroimaging research could become possible.

The sophisticated analysis of neuroimaging data can enable future breakthroughs. Another exciting development is that of real-time fMRI. With this technique, neuroimaging data can be analysed online while the subject is still in the scanner. This analysis can be used to give the subject feedback about their brain activation in real time, so-called neurofeedback. Such feedback can be used to learn how to actively control your brain activation, for example to control your perception of pain.[6] Analysing neuroimaging data online also enables us to control external devices in so-called brain–computer interfaces, which are already being used to help some paralysed patients. How long will it be until we can drive cars using only our thoughts?

Innovative study designs can also help to make neuroimaging studies more realistic. A recent study led by our colleague Paula Banca from the University of Cambridge used live video material while the participants were in the scanner. The aim of the research was to learn more about the neural basis of obsessive–compulsive disorder (OCD). These patients suffer from compulsions (repetitive behaviours the patient feels compelled to carry out) and/or obsessions (intrusive, distressing thoughts that enter the patient's mind). In so-called symptom-provocation designs, researchers try to trigger such obsessions or compulsions to find out what is happening in the patient's brain. Such knowledge is important for a deeper understanding of disorders and to develop new treatments. These obsessions and compulsions differ between individuals—not every patient is afraid of germs or feels compelled to clean excessively. It is therefore important to design these symptom-provocation studies in such a way that they trigger the personal obsessions and compulsions of the patient. Our colleagues developed a creative new design: the researcher went to the patient's home and, with the permission of the patient, performed actions that were tailored to the patient's symptoms. For example, for a patient concerned with symmetry and organization, the researcher would mess up their sock drawer or bed. Live video material of this disorganization was then streamed and shown to the patient, who was undergoing fMRI. Importantly, the patients were free to stop the procedure at any time. The researchers found that the ventromedial prefrontal cortex, an area behind the forehead, acts as the key region in a circuit modulating compulsivity.[7] Such a creative approach can give us new insights into the underlying neural circuits of obsessions, compulsions, and many other conditions.

Thanks to our increasing knowledge about the human brain, we can transform the treatment of neurological and psychiatric disorders. One especially exciting example is deep brain stimulation. In this technique, an electrode which acts like a pacemaker is implanted into the brain. If we know which brain areas function abnormally in a specific disorder, we can sometimes locate them using neuroimaging, then target them and change their activity. Deep brain stimulation is already being used to treat symptoms of Parkinson's disease,[8] clinical depression,[9] and obsessive–compulsive disorder.[10] These patients have failed other lines of treatment, such as medication or behavioural therapy. Often, these cases are very severe, and patients lose hope of returning to a normal life. For example, patients with advanced Parkinson's disease often lose the ability to carry out everyday activities on their own, such as walking or eating. After surgery, we can sometimes see dramatic differences: Parkinson's patients can control their movements again, depressed patients show improvements in mood, and patients with obsessive–compulsive disorder regain control over their thoughts and actions. Such advances would be impossible without detailed knowledge of the underlying brain circuits. With more powerful imaging techniques, new treatments may be discovered. And personalized medicine may become standard practice. If we could find distinct brain abnormalities for each patient with a neurological or psychiatric condition, we could perhaps develop more effective treatments with fewer side effects.

Such advances could be possible thanks to breakthroughs in technology. The fMRI scanners that were used for many studies in this book contain a strong magnet. Its magnetic strength is measured in tesla (T) and the most common strengths are 1.5T and 3T. To give you an idea of the scale: 1.5T is about 30,000 times as strong as the Earth's magnetic field. The stronger the magnet, the better the spatial resolution. While this may already sound strong, some institutions, including the University of Cambridge, are now buying 7T scanners—about 140,000 times the strength of the Earth's magnetic field. What new possibilities and insights will these machines offer? And what ethical dilemmas might result?

While many applications are still science fiction, they might become reality sooner than we imagine. We will need to decide how and where we want to allow mind-reading or neuromarketing. Should we screen for (implicit) racial bias with the help of neuroimaging? Does fMRI-based lie detection violate the privacy of thought? Should we test how moral or self-controlled a person is to prevent future crimes? These are just some of the important ethical questions that will need to be debated and answered by society as a whole, and with the rapid progress of technology, they are becoming increasingly pressing.

NOTES

Chapter 1

1. Owen, A. M. *et al.* Detecting awareness in the vegetative state. *Science* **313**, 1402 (2006).
2. Monti, M. M. *et al.* Willful modulation of brain activity in disorders of consciousness. *N. Engl. J. Med.* **362**, 579–89 (2010).
3. Liu, T. T., Frank, L. R., Wong, E. C., and Buxton, R. B. Detection power, estimation efficiency, and predictability in event-related fMRI. *Neuroimage* **13**, 759–73 (2001).
4. Azevedo, F. A. C. *et al.* Equal numbers of neuronal and nonneuronal cells make the human brain an isometrically scaled-up primate brain. *J. Comp. Neurol.* **513**, 532–41 (2009).
5. McCabe, D. P. and Castel, A. D. Seeing is believing: The effect of brain images on judgments of scientific reasoning. *Cognition* **107**, 343–52 (2008).
6. Michael, R. B., Newman, E. J., Vuorre, M., Cumming, G., and Garry, M. On the (non) persuasive power of a brain image. *Psychon. Bull. Rev.* **20**, 720–5 (2013).

Chapter 2

1. Baron-Cohen, S., Wheelwright, S., Hill, J., Raste, Y., and Plumb, I. The 'Reading the Mind in the Eyes' test revised version: A study with normal adults, and adults with Asperger syndrome or high-functioning autism. *J. Child Psychol. Psychiatry* **42**, 241–51 (2001).
2. Baron-Cohen, S., Leslie, A. M., and Frith, U. Does the autistic child have a 'theory of mind'? *Cognition* **21**, 37–46 (1985).
3. Sample, I. The brain scan that can read people's intentions. *The Guardian* (2007). At <http://www.theguardian.com/science/2007/feb/09/neuroscience.ethicsofscience>, accessed 30 July 2016.
4. Parsons, C. and Waugh, R. So our minds CAN be read: Magnetic scanner produces these actual images from inside people's brains. *Mail Online* (2011).

At <http://www.dailymail.co.uk/sciencetech/article-2040599/Minds-eye-Experts-use-magnetic-scanner-videos-playing-inside-peoples-brains.html>, accessed 30 July 2016.

5. Haynes, J. D. *et al.* Reading hidden intentions in the human brain. *Curr. Biol.* **17**, 323–8 (2007).
6. Soon, C. S., Brass, M., Heinze, H. J., and Haynes, J. D. Unconscious determinants of free decisions in the human brain. *Nat. Neurosci.* **11**, 543–5 (2008).
7. Mitchell, T. M. *et al.* Predicting human brain activity associated with the meanings of nouns. *Science* **320**, 1191–5 (2008).
8. Nishimoto, S. *et al.* Reconstructing visual experiences from brain activity evoked by natural movies. *Curr. Biol.* **21**, 1641–6 (2011).
9. Horikawa, T., Tamaki, M., Miyawaki, Y. and Kamitani, Y. Neural decoding of visual imagery during sleep. *Science* **340**, 639–42 (2013).
10. Kassam, K. S., Markey, A. R., Cherkassky, V. L., Loewenstein, G., and Just, M. A. Identifying emotions on the basis of neural activation. *PLoS One* **8**, e66032 (2013).
11. Eggebrecht, A. T. *et al.* Mapping distributed brain function and networks with diffuse optical tomography. *Nat. Photonics* **8**, 448–54 (2014).

Chapter 3

1. Dovidio, J. F. and Gaertner, S. L. Aversive racism. *Adv. Exp. Soc. Psychol.* **36**, 1–52 (2004).
2. McConahay, J. B. Modern racism, ambivalence, and the Modern Racism Scale, in *Prejudice, discrimination, and racism* (eds. Dovidio, J. F. and Gaertner, S. L.) 91–125 (Academic Press, 1986).
3. Phelps, E. A. *et al.* Performance on indirect measures of race evaluation predicts amygdala activation. *J. Cogn. Neurosci.* **12**, 729–38 (2000).
4. Greenwald, A. G., Mcghee, D. E., and Schwartz, J. L. K. Measuring individual differences in implicit cognition. *J. Pers. Soc. Psychol.* **74**, 1464–80 (1998).
5. Terbeck, S. *et al.* Propranolol reduces implicit negative racial bias. *Psychopharmacology (Berl.)* **222**, 419–24 (2012).
6. Singer, T. *et al.* Empathy for pain involves the affective but not sensory components of pain. *Science* **303**, 1157–62 (2004).
7. Xu, X., Zuo, X., Wang, X., and Han, S. Do you feel my pain? Racial group membership modulates empathic neural responses. *J. Neurosci.* **29**, 8525–9 (2009).
8. Pettigrew, T. F. and Tropp, L. R. A meta-analytic test of intergroup contact theory. *J. Pers. Soc. Psychol.* **90**, 751–83 (2006).

9. Allport, G. W. *The nature of prejudice* (Addison-Wesley Publishing Company, 1979).
10. Zuo, X. and Han, S. Cultural experiences reduce racial bias in neural responses to others' suffering. *Cult. Brain* **1**, 34–46 (2013).
11. Forbes, C. E., Cox, C. L., Schmader, T., and Ryan, L. Negative stereotype activation alters interaction between neural correlates of arousal, inhibition and cognitive control. *Soc. Cogn. Affect. Neurosci.* **7**, 771–81 (2012).
12. Lieberman, M. D., Hariri, A., Jarcho, J. M., Eisenberger, N. I., and Bookheimer, S. Y. An fMRI investigation of race-related amygdala activity in African-American and Caucasian-American individuals. *Nat. Neurosci.* **8**, 720–2 (2005).
13. Hart, A. J. *et al.* Differential response in the human amygdala to racial outgroup vs ingroup face stimuli. *Neuroreport* **11**, 2351–5 (2000).
14. Bertrand, M. and Mullainathan, S. Are Emily and Greg more employable than Lakisha and Jamal? A field experiment on labor market discrimination. *Am. Econ. Rev.* **94**, 991–1013 (2004).
15. Schwartz, L. M., Woloshin, S., and Welch, H. G. Misunderstandings about the effects of race and sex on physicians' referrals for cardiac catheterization. *N. Engl. J. Med.* **341**, 279–83; discussion 286–7 (1999).
16. Korn, H. A., Johnson, M. A., and Chun, M. M. Neurolaw: Differential brain activity for black and white faces predicts damage awards in hypothetical employment discrimination cases. *Soc. Neurosci.* **7**, 398–409 (2012).
17. Wheeler, M. E. and Fiske, S. T. Controlling racial prejudice: Social-cognitive goals affect amygdala and stereotype activation. *Psychol. Sci.* **16**, 56–63 (2005).
18. For example, the University of Cambridge: <http://www.equality.admin.cam.ac.uk/training/equality-diversity-online-training>, accessed 30 July 2016.
19. Hu, X. *et al.* Unlearning implicit social biases during sleep. *Science.* **348**, 1013–15 (2015).

Chapter 4

1. Feldman, R. S., Forrest, J. A., and Happ, B. R. Self-presentation and verbal deception: Do self-presenters lie more? *Basic Appl. Soc. Psych.* **24**, 163–70 (2002).
2. Serota, K. B., Levine, T. R., and Boster, F. J. The prevalence of lying in America: Three studies of self-reported lies. *Hum. Commun. Res.* **36**, 2–25 (2010).
3. Debey, E., de Schryver, M., Logan, G. D., Suchotzki, K., and Verschuere, B. From junior to senior Pinocchio: A cross-sectional lifespan investigation of deception. *Acta Psychol. (Amst).* **160**, 58–68 (2015).

4. Bond, C. F. and DePaulo, B. M. Accuracy of deception judgments. *Pers. Soc. Psychol. Rev.* **10**, 214–34 (2006).
5. Trovillo, P. V. A history of lie detection. *J. Crim. Law Criminol.* **29**, 848 (1939).
6. Brett, A. S., Phillips, M., and Beary, J. F. Predictive power of the polygraph: Can the 'lie detector' really detect liars? *Lancet* **1**, 544–7 (1986).
7. No Lie MRI website at <http://www.noliemri.com/>, accessed 30 July 2016.
8. Natu, N. This brain test maps the truth. *The Times of India* (2008). At <http://timesofindia.indiatimes.com/city/mumbai/This-brain-test-maps-the-truth/articleshow/3257032.cms?>, accessed 30 July 2016.
9. Murphy, E. Update on Indian BEOS case: Accused released on bail. *Stanford Center for Law & the Biosciences Blog* (2009). At <https://lawandbiosciences.wordpress.com/2009/04/02/update-on-indian-beos-case-accused-released-on-bail/>, accessed 30 July 2016.
10. Murphy, E. No Lie MRI being offered as evidence in court. *Stanford Center for Law & the Biosciences Blog* (2009). At <http://blogs.law.stanford.edu/lawandbiosciences/2009/03/14/no-lie-mri-being-offered-as-evidence-in-court/>, accessed 30 July 2016.
11. Washburn, D. Can this machine prove if you're lying? *Voice of San Diego* (2009). At <http://www.voiceofsandiego.org/science/can-this-machine-prove-if-youre-lying/>, accessed 30 July 2016.
12. Expert Testimony Judge Tu M. Pham in the case *USA v Semrau.* (2010). At <https://www.tnwd.uscourts.gov/JudgePham/opinions/659.pdf>, accessed 30 July 2016.
13. Spence, S. A. *et al.* Behavioural and functional anatomical correlates of deception in humans. *Neuroreport* **12**, 2849–53 (2001).
14. Levy, B. J. and Wagner, A. D. Cognitive control and right ventrolateral prefrontal cortex: Reflexive reorienting, motor inhibition, and action updating. *Ann. N. Y. Acad. Sci.* **1224**, 40–62 (2011).
15. de Carli, D. *et al.* Identification of activated regions during a language task. *Magn. Reson. Imaging* **25**, 933–8 (2007).
16. Demb, J. B. *et al.* Semantic encoding and retrieval in the left inferior prefrontal cortex: A functional MRI study of task difficulty and process specificity. *J. Neurosci.* **15**, 5870–8 (1995).
17. Brunet, E., Sarfati, Y., Hardy-Baylé, M. C., and Decety, J. A PET investigation of the attribution of intentions with a nonverbal task. *Neuroimage* **11**, 157–66 (2000).
18. Rogers, R. D. *et al.* Choosing between small, likely rewards and large, unlikely rewards activates inferior and orbital prefrontal cortex. *J. Neurosci.* **19**, 9029–38 (1999).

19. Langleben, D. D. *et al.* Brain activity during simulated deception: An event-related functional magnetic resonance study. *Neuroimage* **15**, 727–32 (2002).
20. Kerns, J. G. *et al.* Anterior cingulate conflict monitoring and adjustments in control. *Science* **303**, 1023–6 (2004).
21. Beauregard, M., Lévesque, J., and Bourgouin, P. Neural correlates of conscious self-regulation of emotion. *J. Neurosci.* **21**, RC165 (2001).
22. Hakun, J. G. *et al.* fMRI investigation of the cognitive structure of the Concealed Information Test. *Neurocase: The Neural Basis of Cognition.* **14**, 59–67 (2008).
23. Langleben, D. D. *et al.* Telling truth from lie in individual subjects with fast event-related fMRI. *Hum. Brain Mapp.* **26**, 262–72 (2005).
24. Davatzikos, C. *et al.* Classifying spatial patterns of brain activity with machine learning methods: Application to lie detection. *Neuroimage* **28**, 663–8 (2005).
25. Kozel, F. A. *et al.* Detecting deception using functional magnetic resonance imaging. *Biol. Psychiatry* **58**, 605–13 (2005).
26. Ganis, G., Rosenfeld, J. P., Meixner, J., Kievit, R. A., and Schendan, H. E. Lying in the scanner: Covert countermeasures disrupt deception detection by functional magnetic resonance imaging. *Neuroimage* **55**, 312–19 (2011).
27. Farah, M. J., Hutchinson, J. B., Phelps, E. A., and Wagner, A. D. Functional MRI-based lie detection: Scientific and societal challenges. *Nat. Publ. Gr.* **15**, 123–31 (2014).
28. Jiang, W. *et al.* A functional MRI study of deception among offenders with antisocial personality disorders. *Neuroscience* **244**, 90–8 (2013).
29. Ganis, G., Kosslyn, S. M., Stose, S., Thompson, W. L., and Yurgelun-Todd, D. A. Neural correlates of different types of deception: An fMRI investigation. *Cereb. Cortex* **13**, 830–6 (2003).
30. Weisberg, D. S., Keil, F. C., Goodstein, J., Rawson, E., and Gray, R. The seductive allure of neuroscience explanations. *J. Cogn. Neurosci.* **20**, 470–7 (2008).

Chapter 5

1. Thomson, J. J. The trolley problem. *Yale Law J.* **94**, 1395–415 (1985).
2. Hauser, M., Cushman, F., Young, L., Jin, R. K.-X., and Mikhail, J. A dissociation between moral judgments and justifications. *Mind & Language* **22**, 1–21 (2007).
3. Greene, J. D., Sommerville, R. B., Nystrom, L. E., Darley, J. M., and Cohen, J. D. An fMRI investigation of emotional engagement in moral judgment. *Science* **293**, 2105–8 (2001).

4. Ciaramelli, E., Muccioli, M., Làdavas, E., and di Pellegrino, G. Selective deficit in personal moral judgment following damage to ventromedial prefrontal cortex. *Soc. Cogn. Affect. Neurosci.* 84–92 (2007).
5. Young, L. and Saxe, R. Innocent intentions: A correlation between forgiveness for accidental harm and neural activity. *Neuropsychologia* **47**, 2065–72 (2009).
6. Saxe, R. and Kanwisher, N. People thinking about thinking people: The role of the temporo-parietal junction in 'theory of mind'. *Neuroimage* **19**, 1835–42 (2003).
7. Moll, J., Eslinger, P. J., and de Oliveira-Souza, R. Frontopolar and anterior temporal cortex activation in a moral judgment task: Preliminary functional MRI results in normal subjects. *Arq. Neuropsiquiatr.* **59**, 657–64 (2001).
8. Moll, J., de Oliveira-Souza, R., Bramati, I. E., and Grafman, J. Functional networks in emotional moral and nonmoral social judgments. *Neuroimage* **16**, 696–703 (2002).
9. Moll, J. *et al.* The neural correlates of moral sensitivity: A functional magnetic resonance imaging investigation of basic and moral emotions. *J. Neurosci.* **22**, 2730–6 (2002).
10. Hein, G. and Knight, R. T. Superior temporal sulcus: It's my area: or is it? *J. Cogn. Neurosci.* **20**, 2125–36 (2008).
11. Williamson, S., Hare, R. D., and Wong, S. Violence: Criminal psychopaths and their victims. *Can. J. Behav. Sci.* **19**, 454–62 (1987).
12. Meffert, H., Gazzola, V., den Boer, J. A., Bartels, A. A. J., and Keysers, C. Reduced spontaneous but relatively normal deliberate vicarious representations in psychopathy. *Brain* **136**, 2550–62 (2013).
13. Blair, R. J. R. A cognitive developmental approach to morality: Investigating the psychopath. *Cognition* **57**, 1–29 (1995).
14. Blair, R. J. R. Moral reasoning and the child with psychopathic tendencies. *Pers. Individ. Dif.* **22**, 731–9 (1997).
15. Harenski, C. L., Harenski, K. A., Shane, M. S., and Kiehl, K. A. Aberrant neural processing of moral violations in criminal psychopaths. *J. Abnorm. Psychol.* **119**, 863–74 (2010).
16. Coid, J., Yang, M., Ullrich, S., Roberts, A., and Hare, R. D. Prevalence and correlates of psychopathic traits in the household population of Great Britain. *Int. J. Law Psychiatry* **32**, 65–73 (2009).
17. Cooke, D. J. Psychopathic personality in different cultures: What do we know? What do we need to find out? *J. Pers. Disord.* **10**, 23–40 (1996).
18. Coid, J. *et al.* Psychopathy among prisoners in England and Wales. *Int. J. Law Psychiatry* **32**, 134–41 (2009).

19. Ullrich, S., Paelecke, M., Kahle, I., and Marneros, A. Kategoriale und dimensionale Erfassung von 'psychopathy' bei deutschen Straftätern. Prävalenz, Geschlechts- und Alterseffekte. *Nervenarzt* **74**, 1002–8 (2003).
20. Assadi, S. M. *et al.* Psychiatric morbidity among sentenced prisoners: Prevalence study in Iran. *Br. J. Psychiatry* **188**, 159–64 (2006).
21. Hughes, V. Science in court: Head case. *Nature* **464**, 340–2 (2010).
22. Zorn, E. Passing thought—Today truly marks the end of the Nicarico murder case. *Chicago Tribune* (2011). At <http://blogs.chicagotribune.com/news_columnists_ezorn/2011/03/passing-thought-today-truly-marks-the-end-of-the-nicarico-murder-case.html>, accessed 30 July 2016.
23. O'Reardon, J. P. *et al.* Efficacy and safety of transcranial magnetic stimulation in the acute treatment of major depression: A multisite randomized controlled trial. *Biol. Psychiatry* **62**, 1208–16 (2007).
24. Talelli, P., Greenwood, R. J., and Rothwell, J. C. Exploring Theta Burst Stimulation as an intervention to improve motor recovery in chronic stroke. *Clin. Neurophysiol.* **188**, 333–42 (2007).
25. Aleman, A., Sommer, I. E. C., and Kahn, R. S. Efficacy of slow repetitive transcranial magnetic stimulation in the treatment of resistant auditory hallucinations in schizophrenia: A meta-analysis. *J. Clin. Psychiatry* **68**, 416–21 (2007).
26. Young, L., Camprodon, J. A., Hauser, M., Pascual-Leone, A., and Saxe, R. Disruption of the right temporoparietal junction with transcranial magnetic stimulation reduces the role of beliefs in moral judgments. *Proc. Natl. Acad. Sci. USA* **107**, 6753–8 (2010).
27. Wheatley, T. and Haidt, J. Hypnotic disgust makes moral judgments more severe. **16**, 1–6 (2005).
28. Crockett, M. J. *et al.* Harm to others outweighs harm to self in moral decision making. *Proc. Natl. Acad. Sci.* **111**, 17320–5 (2014).
29. Crockett, M. J. *et al.* Dissociable effects of serotonin and dopamine on the valuation of harm in moral decision making. *Curr. Biol.* **25**, 1852–9 (2015).
30. Merims, D. and Giladi, N. Dopamine dysregulation syndrome, addiction and behavioral changes in Parkinson's disease. *Park. Relat. Disord.* **14**, 273–80 (2008).

Chapter 6

1. Mischel, W., Ebbesen, E. B., and Zeiss, A. R. Cognitive and attentional mechanisms in delay of gratification. *J. Pers. Soc. Psychol.* **21**, 204–18 (1972).
2. Mischel, W., Shoda, Y., and Peake, P. K. The nature of adolescent competencies predicted by preschool delay of gratification. *J. Pers. Soc. Psychol.* **54**, 687–96 (1988).

3. Shoda, Y., Mischel, W., and Peake, P. K. Predicting adolescent cognitive and self-regulatory competencies from preschool delay of gratification: Identifying diagnostic conditions. *Dev. Psychol.* **26**, 978–86 (1990).
4. Schlam, T. R., Wilson, N. L., Shoda, Y., Mischel, W., and Ayduk, O. Preschoolers' delay of gratification predicts their body mass 30 years later. *J. Pediatr.* **162**, 90–3 (2013).
5. Kidd, C., Palmeri, H., and Aslin, R. N. Rational snacking: Young children's decision-making on the marshmallow task is moderated by beliefs about environmental reliability. *Cognition* **126**, 109–14 (2013).
6. Ayduk, O. *et al.* Regulating the interpersonal self: Strategic self-regulation for coping with rejection sensitivity. *J. Pers. Soc. Psychol.* **79**, 776–92 (2000).
7. Stroop, J. R. Studies of interference in serial verbal reactions. *J. Exp. Psychol.* **18**, 643–62 (1935).
8. Marsh, R. *et al.* A developmental fMRI study of self-regulatory control. *Hum. Brain Mapp.* **27**, 848–63 (2006).
9. Chudasama, Y. and Robbins, T. W. Functions of frontostriatal systems in cognition: Comparative neuropsychopharmacological studies in rats, monkeys and humans. *Biol. Psychol.* **73**, 19–38 (2006).
10. Konishi, S., Nakajima, K., Uchida, I., Sekihara, K., and Miyashita, Y. No-go dominant brain activity in human inferior prefrontal cortex revealed by functional magnetic resonance imaging. *Eur. J. Neurosci.* **10**, 1209–13 (1998).
11. Aron, A. R., Fletcher, P. C., Bullmore, E. T., Sahakian, B. J., and Robbins, T. W. Stop-signal inhibition disrupted by damage to right inferior frontal gyrus in humans. *Nat. Neurosci.* **6**, 115–16 (2003).
12. Nee, D. E., Wager, T. D., and Jonides, J. Interference resolution: Insights from a meta-analysis of neuroimaging tasks. *Cogn. Affect. Behav. Neurosci.* **7**, 1–17 (2007).
13. McClure, S. M., Laibson, D. I., Loewenstein, G., and Cohen, J. D. Separate neural systems value immediate and delayed monetary rewards. *Science* **306**, 503–7 (2004).
14. Knutson, B., Fong, G. W., Adams, C. M., Varner, J. L., and Hommer, D. Dissociation of reward anticipation and outcome with event-related fMRI. *Neuroreport* **12**, 3683–7 (2001).
15. Beauregard, M., Lévesque, J., and Bourgouin, P. Neural correlates of conscious self-regulation of emotion. *J. Neurosci.* **21**, RC165 (2001).
16. Devinsky, O., Morrell, M. J., and Vogt, B. A. Contributions of anterior cingulate cortex to behaviour. *Brain* **118, Pt 1,** 279–306 (1995).
17. Holroyd, C. B. and Coles, M. G. H. The neural basis of human error processing: Reinforcement learning, dopamine, and the error-related negativity. *Psychol. Rev.* **109**, 679–709 (2002).

18. Shilling, V. M., Chetwynd, A., and Rabbitt, P. M. A. Individual inconsistency across measures of inhibition: An investigation of the construct validity of inhibition in older adults. *Neuropsychologia* **40**, 605–19 (2002).
19. Menzies, L. *et al.* Integrating evidence from neuroimaging and neuropsychological studies of obsessive–compulsive disorder: The orbitofronto-striatal model revisited. *Neurosci. Biobehav. Rev.* **32**, 525–49 (2008).
20. Leckman, J. F., Bloch, M. H., Smith, M. E., Larabi, D., and Hampson, M. Neurobiological substrates of Tourette's disorder. *J. Child Adolesc. Psychopharmacol.* **20**, 237–47 (2010).
21. Cubillo, A., Halari, R., Smith, A., Taylor, E., and Rubia, K. A review of fronto-striatal and fronto-cortical brain abnormalities in children and adults with Attention Deficit Hyperactivity Disorder (ADHD) and new evidence for dysfunction in adults with ADHD during motivation and attention. *Cortex* **48**, 194–215 (2012).
22. Grafman, J. *et al.* Frontal lobe injuries, violence, and aggression: A report of the Vietnam Head Injury Study. *Neurology* **46**, 1231–8 (1996).
23. Burns, J. M. and Swerdlow, R. H. Right orbitofrontal tumor with pedophilia symptom and constructional apraxia sign. *Arch. Neurol.* **60**, 437–40 (2003).
24. Moll, J. *et al.* The neural correlates of moral sensitivity: A functional magnetic resonance imaging investigation of basic and moral emotions. *J. Neurosci.* **22**, 2730–6 (2002).
25. Gottfredson, M. R. and Hirschi, T. A *general theory of crime* (Stanford University Press, 1990).
26. Pratt, T. C. and Cullen, F. T. The empirical status of Gottfredson and Hirschi's general theory of crime: A meta-analysis. *Criminology* **38**, 931–64 (2000).
27. For a good critical review, see: Geis, G. On the absence of self-control as the basis for a General Theory of Crime: A critique. *Theor. Criminol.* **4**, 35–53 (2000).
28. Aharoni, E. *et al.* Neuroprediction of future rearrest. *Proc. Natl Acad. Sci. USA* **110**, 6223–8 (2013).
29. Baumeister, R. F. and Heatherton, T. F. Self-regulation failure: An overview. *Psychol. Inq.* **7**, 1–15 (1996).
30. Baumeister, R. F., Bratslavsky, E., Muraven, M., and Tice, D. M. Ego depletion: Is the active self a limited resource? *J. Pers. Soc. Psychol.* **74**, 1252–65 (1998).
31. Hagger, M. S., Wood, C., Stiff, C., and Chatzisarantis, N. L. D. Ego depletion and the strength model of self-control: A meta-analysis. *Psychol. Bull.* **136**, 495–525 (2010).

32. Sripada, C., Kessler, D., and Jonides, J. Methylphenidate blocks effort-induced depletion of regulatory control in healthy volunteers. *Psychol. Sci.* **25**, 1227–34 (2014).
33. Hagger, M. S. *et al.* A multi-lab pre-registered replication of the ego-depletion effect. *Perspect. Psychol. Sci.* (in press).
34. Baumeister, R. F. and Vohs, K. D. Misguided effort with elusive implications. *Perspect. Psychol. Sci.* (Epub ahead of print).
35. Inzlicht, M. and Schmeichel, B. J. What is ego depletion? Toward a mechanistic revision of the resource model of self-control. *Perspect. Psychol. Sci.* **7**, 450–63 (2012).
36. Muraven, M. and Slessareva, E. Mechanisms of self-control failure: Motivation and limited resources. *Pers. Soc. Psychol. Bull.* **29**, 894–906 (2003).
37. Heatherton, T. F. and Wagner, D. D. Cognitive neuroscience of self-regulation failure. *Trends Cogn. Sci.* **15**, 132–9 (2011).
38. Volkow, N. D. *et al.* Cognitive control of drug craving inhibits brain reward regions in cocaine abusers. *Neuroimage* **49**, 2536–43 (2010).
39. Wagner, D. D., Altman, M., Boswell, R. G., Kelley, W. M., and Heatherton, T. F. Self-regulatory depletion enhances neural responses to rewards and impairs top-down control. *Psychol. Sci.* **24**, 2262–71 (2013).
40. Maier, S. U., Makwana, A. B., and Hare, T. A. Acute stress impairs self-control in goal-directed choice by altering multiple functional connections within the brain's decision circuits. *Neuron* **87**, 621–31 (2015).
41. Adam, T. C. and Epel, E. S. Stress, eating and the reward system. *Physiol. Behav.* **91**, 449–58 (2007).
42. Everitt, B. J., Cador, M., and Robbins, T. W. Interactions between the amygdala and ventral striatum in stimulus–reward associations: Studies using a second-order schedule of sexual reinforcement. *Neuroscience* **30**, 63–75 (1989).
43. Cardinal, R. N., Parkinson, J. A., Hall, J., and Everitt, B. J. Emotion and motivation: The role of the amygdala, ventral striatum, and prefrontal cortex. *Neurosci. Biobehav. Rev.* **26**, 321–52 (2002).
44. Daniel, T. O., Stanton, C. M., and Epstein, L. H. The future is now: Reducing impulsivity and energy intake using episodic future thinking. *Psychol. Sci.* **24**, 2339–42 (2013).
45. Muraven, M. Building self-control strength: Practicing self-control leads to improved self-control performance. *J. Exp. Soc. Psychol.* **46**, 465–8 (2010).
46. Muraven, M. Practicing self-control lowers the risk of smoking lapse. *Psychol. Addict. Behav.* **24**, 446–52 (2010).

47. Soon, C. S., Brass, M., Heinze, H.-J., and Haynes, J. D. Unconscious determinants of free decisions in the human brain. *Nat. Neurosci.* **11**, 543–5 (2008).
48. Fried, I., Mukamel, R., and Kreiman, G. Internally generated preactivation of single neurons in human medial frontal cortex predicts volition. *Neuron* **69**, 548–62 (2011).
49. Tang, Y.-Y., Posner, M. I., Rothbart, M. K., and Volkow, N. D. Circuitry of self-control and its role in reducing addiction. *Trends Cogn. Sci.* **19**, 439–44 (2015).

Chapter 7

1. The calculation of ad success is based on the so-called advertising elasticity, which depends on a variety of variables and basically describes how much sales will increase by spending 1 per cent more for advertising.

2. Venkatraman, V. *et al.* Predicting advertising success beyond traditional measures: New insights from neurophysiological methods and market response modeling. *J. Mark. Res.* **LII**, 436–52 (2015).
3. O'Doherty, J. *et al.* Dissociable roles of ventral and dorsal striatum in instrumental conditioning. *Science* **304**, 452–4 (2004).
4. McClure, S. M. *et al.* Neural correlates of behavioral preference for culturally familiar drinks. *Neuron* **44**, 379–87 (2004).
5. Kable, J. W. and Glimcher, P. W. The neural correlates of subjective value during intertemporal choice. *Nat. Neurosci.* **10**, 1625–33 (2007).
6. Cabeza, R. and Nyberg, L. Imaging cognition II: An empirical review of 275 PET and fMRI studies. *J. Cogn. Neurosci.* **12**, 1–47 (2000).
7. Nee, D. E., Wager, T. D., and Jonides, J. Interference resolution: Insights from a meta-analysis of neuroimaging tasks. *Cogn. Affect. Behav. Neurosci.* **7**, 1–17 (2007).
8. Plassmann, H., Doherty, J. O., Shiv, B., and Rangel, A. Marketing actions can modulate neural representations of experienced pleasantness. *Proc. Nationa Acad. Sci. USA* **105**, (2008).
9. Ishizu, T. and Zeki, S. Toward a brain-based theory of beauty. *PLoS One* **6**, e21852 (2011).
10. Berns, G. S. and Moore, S. E. A neural predictor of cultural popularity. *J. Consum. Psychol.* **22**, 154–60 (2012).
11. Knutson, B., Adams, C. M., Fong, G. W., and Hommer, D. Anticipation of increasing monetary reward selectively recruits nucleus accumbens. *J. Neurosci.* **21**, RC159 (2001).

12. Yoon, C., Gutchess, A. H., Feinberg, F., and Polk, T. A. A functional magnetic resonance imaging study of neural dissociations between brand and person judgments. *J. Consum. Res.* **33**, 31–40 (2006).
13. Poldrack, R. A. Can cognitive processes be inferred from neuroimaging data? *Trends Cogn. Sci.* **10**, 59–63 (2006).
14. Lindstrom, M. You love your iPhone. Literally. *The New York Times* (2011). At <http://www.nytimes.com/2011/10/01/opinion/you-love-your-iphone-literally.html>, accessed 31 July 2016.
15. Yarkoni, T., Poldrack, R. A., Nichols, T. E., Van Essen, D. C., and Wager, T. D. Large-scale automated synthesis of human functional neuroimaging data. *Nat. Methods* **8**, 665–70 (2011).
16. Poldrack, R. The iPhone and the brain. *The New York Times* (2011). At <http://www.nytimes.com/2011/10/05/opinion/the-iphone-and-the-brain.html>, accessed 31 July 2016.
17. US Food and Drug Administration. MRI (Magnetic Resonance Imaging). At <http://www.fda.gov/Radiation-EmittingProducts/RadiationEmittingProductsandProcedures/MedicalImaging/MRI/default.htm>, accessed 10 August 2016.
18. Illes, J. *et al.* Discovery and disclosure of incidental findings in neuroimaging research. *J. Magn. Reson. Imaging* **20**, 743–7 (2004).
19. Wolf, S. M. Incidental findings in neuroscience research: A fundamental challenge to the structures of bioethics and health law, in *The Oxford Handbook of Neuroethics* (eds. Illes, J. and Sahakian, B. J.) 623–34 (OUP, 2011).
20. Booth, T. C., Jackson, A., Wardlaw, J. M., Taylor, S. A., and Waldman, A. D. Incidental findings found in 'healthy' volunteers during imaging performed for research: Current legal and ethical implications. *Br. J. Radiol.* **83**, 456–65 (2010).

Chapter 8

1. Roiser, J. P. and Sahakian, B. J. Hot and cold cognition in depression. *CNS Spectr.* **18**, 139–49 (2013).
2. Giedd, J. N. *et al.* Brain development during childhood and adolescence: A longitudinal MRI study. *Nat. Neurosci.* **2**, 861–3 (1999).
3. Gogtay, N. *et al.* Dynamic mapping of human cortical development during childhood through early adulthood. *Proc. Natl Acad. Sci. USA* **101**, 8174–9 (2004).

4. Jernigan, T. L. *et al.* The pediatric imaging, neurocognition, and genetics (PING) data repository. *Neuroimage* **124**, 1149–54 (2016).
5. Gibbs, S. Google buys UK artificial intelligence startup Deepmind for £400m. *The Guardian* (2014). At <http://www.theguardian.com/technology/2014/jan/27/google-acquires-uk-artificial-intelligence-startup-deepmind>, accessed 31 July 2016.
6. deCharms, R. C. *et al.* Control over brain activation and pain learned by using real-time functional MRI. *Proc. Natl Acad. Sci. USA* **102**, 18626–31 (2005).
7. Banca, P. *et al.* Imbalance in habitual versus goal directed neural systems during symptom provocation in obsessive–compulsive disorder. *Brain* **138**, 798–811 (2015).
8. Krack, P. *et al.* Five-year follow-up of bilateral stimulation of the subthalamic nucleus in advanced Parkinson's disease. *N. Engl. J. Med.* **349**, 1925–34 (2003).
9. Mayberg, H. S. *et al.* Deep brain stimulation for treatment-resistant depression. *Neuron* **45**, 651–60 (2005).
10. Abelson, J. L. *et al.* Deep brain stimulation for refractory obsessive–compulsive disorder. *Biol. Psychiatry* **57**, 510–16 (2005).

INDEX

Figures are indicated by an italic *f* following the page number.

BAD MOVES

How decision making goes wrong, and the ethics of smart drugs

Barbara Sahakian and Jamie Nicole LaBuzetta

978-0-19-966848-9 | Paperback | £8.99

'With this accessible primer, full of medical anecdotes and clear explanations, Sahakian and LaBuzetta prepare the public for an informed discussion about the role of drugs in our society.'

Nature

The realization that smart drugs can improve cognitive abilities in healthy people has led to growing general use, with drugs easily available via the Internet. Sahakian and LaBuzetta raise ethical questions about the availability of these drugs for cognitive enhancement, in the hope of informing public debate about an increasingly important issue.

THE BRAIN SUPREMACY

Notes from the frontiers of neuroscience

Kathleen Taylor

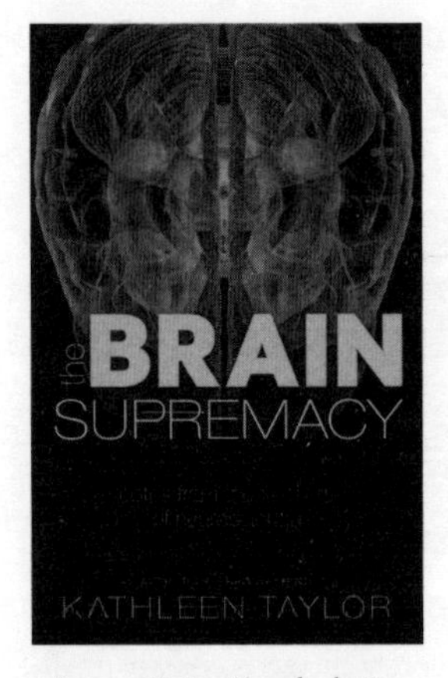

978-0-19-968385-7 | Paperback | £12.99

'This is a thoughtful guide to a rapidly advancing field and its implications.'

Network Review

In recent years funding and effort have been poured into brain research. We are entering the era of the brain supremacy. What will the new science mean for us, as individuals, consumers, parents, and citizens? Should we be excited, or alarmed, by the remarkable promises of drugs that can boost our brain power, ever more subtle marketing techniques, and even machines that can read minds?

The Brain Supremacy is a lucid and rational guide to this exciting new world. Using recent examples from the scientific literature and the media, it explores the science behind the hype. Looking to the future, this book sets current neuroscience in its social and ethical context, as an increasingly important influence on how all of us live our lives.

IGNORANCE

How It Drives Science

Stuart Firestein

978-0-19-982807-4 | Hardback | £15.49

'Packed with real examples and deep practical knowledge, *Ignorance* is a thoughtful introduction to the nature of knowing, and the joy of curiosity.'

Adam Rutherford, *The Observer*

Most of us have a false impression of science as a sure-fire, deliberate, step-by-step method for finding things out and getting things done. In fact, says Firestein, more often than not, science is like looking for a black cat in a dark room, and there may not be a cat in the room. But it is exactly this 'not knowing', this puzzling over thorny questions or inexplicable data, that gets researchers into the lab early and keeps them there late, the thing that propels them, the very driving force of science. In his book, Firestein shows how scientists use ignorance to program their work, to identify what should be done, what the next steps are, and where they should concentrate their energies, opening a new window on the true nature of research.

SOCIAL

Why our brains are wired to connect

Matthew D. Lieberman

978-0-19-874381-1 | Paperback | £12.99

'This isn't just fascinating for its own sake. Lieberman has a social and political purpose.'

Julian Baggini, *Financial Times*

'*Social* is the book I've been waiting for: a brilliant and beautiful exploration of how and why we are wired together, by one of the field's most prescient pioneers.'

Daniel Gilbert, Harvard University

Why are we influenced by the behaviour of complete strangers? Why does the brain register similar pleasure when I perceive something as 'fair' as when I eat chocolate? Why can we be so profoundly hurt by bereavement? The young discipline of 'social cognitive neuroscience' has been exploring this fascinating interface between brain science and human behaviour since the late 1990s.